ROADSIDE GEOLOGY

of South Dakota

John Paul Gries

Mountain Press Publishing Company
Missoula, Montana

Library of Congress Cataloging-in-Publication Data

Gries, John Paul, 1911-
    Roadside geology of South Dakota / John Paul Gries.
        p.   cm.
    Includes bibliographical references (p.   -   ) and index.
    ISBN 0-87842-338-9 (pbk.   :   alk. paper)
    1. Geology—South Dakota—Guidebooks.   2. South Dakota—
Guidebooks.   I. Title.
QE163.G68   1996
557.83—dc20   96-15930
                                                                        CIP

Printed in United States of America

MOUNTAIN PRESS PUBLISHING COMPANY
P.O. Box 2399 • Missoula, MT 59806
406-728-1900 • 800-234-5308

To the many field geologists who, over the past 120 years have covered the state on foot, on horseback, and more recently by automobile, carefully observing, recording, and interpreting the details of the surface geology of South Dakota.

My model is Nathaniel Horatio Darton, geologist with the United States Geological Survey, who in the early 1900s began studying the underground aquifers of the state. He later prepared maps of the surface geology of the Black Hills and adjacent prairies, which I still use nearly every day in answering inquiries about the availability and depth to groundwater in the western half of South Dakota. Even in this day of gadgetry, field studies are still the base upon which the modern geologist must build.

Roads and highways of South Dakota.

Contents

Preface

I wrote this book for the people of South Dakota and for visitors to the state who would like to know more about the rocks they see about them. I intended it for people who are not geologists and made every effort to address them in their own terms.

The book begins with a general introduction to the rocks of South Dakota. Then it breaks the state down into three large parts: the areas east and west of the Missouri River and the Black Hills. Each of those chapters begins with a more detailed introduction, then continues with a series of road guides, which provide details of what you see along the way.

South-to-north road guides are written as continuous descriptions of the 200-mile width of the state. East-to-west road guides break at the Missouri River. Natives refer to the eastern section as "East River" and to the area west of the Missouri River as "West River." The first section of this guide covers the East River counties, the second the West River counties, and the third, the Black Hills area.

An overview precedes each road log as a brief description of what is to come. Short segments then divide the log, which may be read either way, depending upon the direction of travel.

Acknowledgments

This book would not have been possible without the help of many professional colleagues; to each I am deeply grateful. Bill Roggenthen, department chair, has encouraged continued use of office space and other facilities of the Department of Geology and Geological Engineering at South Dakota School of Mines and Technology. Professors Jim Fox, Alvis Lisenbee, Colin Paterson, and Jack Redden have helped on a daily basis with discussions, suggestions, and review of many parts of the text and illustrations.

Merlin Tipton, former director of the South Dakota Geological Survey at Vermillion, gave access to published and unpublished material in their files and made available their extensive photo collection. I am particularly grateful for permission to use sections of the still-unpublished glacial map of eastern South Dakota. At the Survey's Rapid City office, Fred Steece and Lynn Hedges have shared with me their knowledge of many years of field mapping the glaciated area east of the Missouri River.

I have been in almost daily touch with Phil Bjork, director of the Museum of Geology at the School of Mines. In addition to answering countless questions on the geology and paleontology of the Badlands, he furnished many photographs taken in the field, and in the museum. I especially thank Joyce Tripp, who had the patience to transcribe my manually typed manuscript onto a floppy disk.

James Martin, of both the South Dakota Geological Survey and the Museum of Geology, helped with the glacial geology east of the Missouri River and with the Tertiary geology of the Badlands in the southwest quadrant of the state.

I made free use of the many publications of the U. S. Geological Survey, and used their geologic map of South Dakota (1951) as the basis for the strip highway maps west of the Missouri River. The newer geologic map of the Black Hills (1989), by Redden and others, was the base for the detailed strip maps in that region.

Driving the highways to make or check the logs was an educational experience for me, but probably a bit challenging for my companions. I particularly acknowledge Lenord Yarger of Rapid City for his company and help on these expeditions.

I much appreciate the base maps for the geologic road logs from Vernon Bump of the South Dakota Department of Transportation.

Photo credit is cited for individuals, companies, and organizations. Steve Baldwin, Art Palmer, and Mike Wiles furnished illustrative material for discussions of Jewel and Wind Caves. They spent hundreds of hours underground, mapping and studying the origin of limestone caverns.

I especially appreciate my wife, Virginia, who has foregone many of the pleasures of our retirement to allow time for the preparation of *Roadside Geology of South Dakota*.

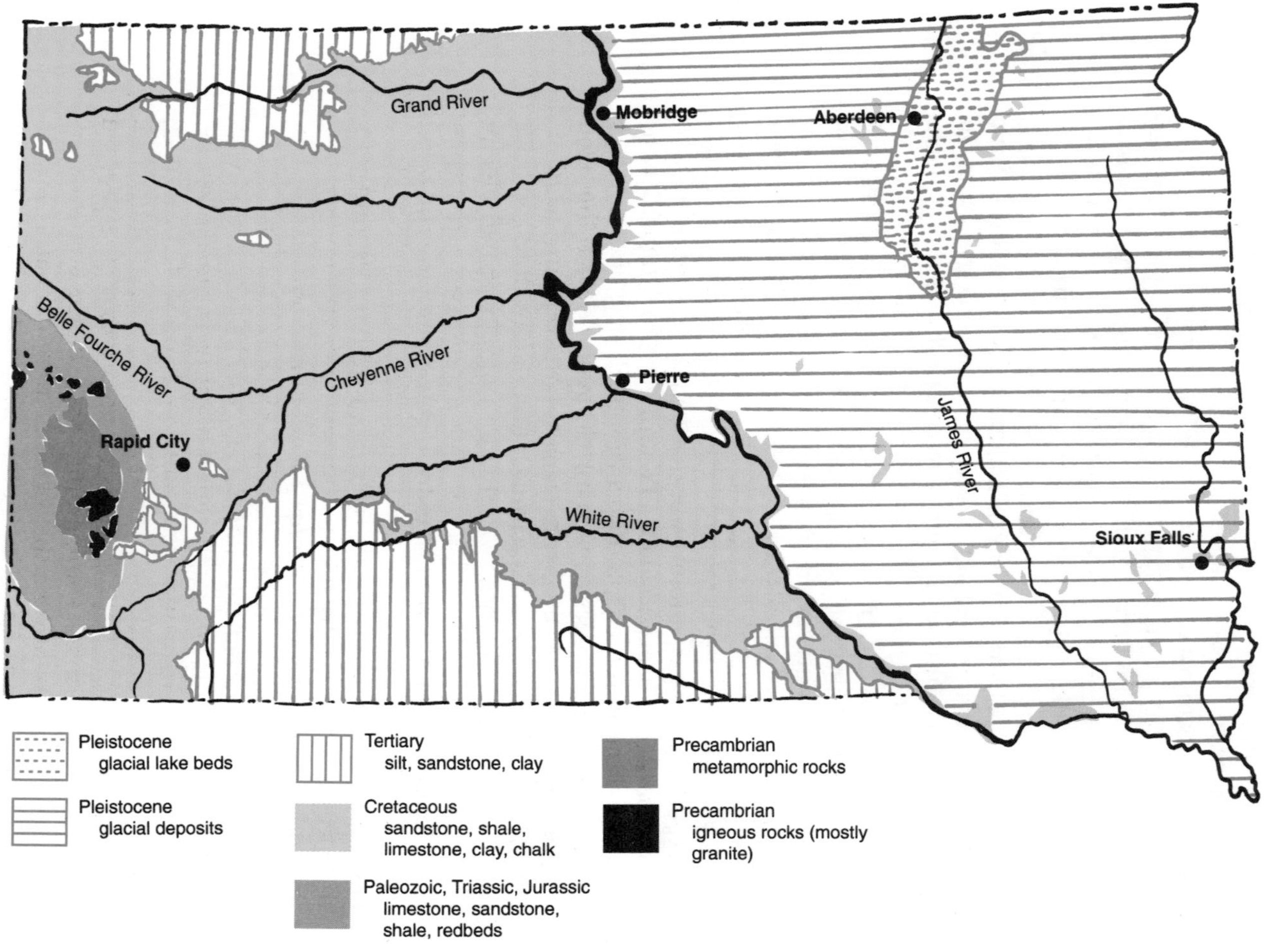

General geology of South Dakota.

Precambrian metamorphic rocks

Precambrian igneous rocks (mostly granite)

Tertiary silt, sandstone, clay

Cretaceous sandstone, shale, limestone, clay, chalk

Paleozoic, Triassic, Jurassic limestone, sandstone, shale, redbeds

Pleistocene glacial lake beds

Pleistocene glacial deposits

Sioux Falls

James River

Aberdeen

Mobridge

Pierre

White River

Grand River

Cheyenne River

Belle Fourche River

Rapid City

Introduction

A drive across South Dakota reveals a variety of landscapes, from the spectacular granite spires of the higher Black Hills and the concentric hogback ridges that encircle them to the precipitous Badlands, the rolling prairies, and the deep trench of the Missouri River. Farther east are the more subtle landforms left by the melting glaciers of 10,000 or so years ago.

Geographically and economically, the Missouri River divides the state. The glaciated area east of the Missouri River is gently rolling, well drained, and receives enough rainfall for farming, dairying, and some intensive livestock raising. To the west, the land is rough, the soil thin, rainfall sparse. Ranching and dry farming are the dominant industries. The forested Black Hills offer lumbering and summer grazing, and mines there produce a variety of rocks and minerals.

The rocks and landscapes of South Dakota correlate rather well with the nationalities of the immigrants they attracted. The glaciated land east of the Missouri River, with fertile soil, relatively smooth topography, and adequate rainfall, attracted English, Scandinavian, and German farmers in the 1870s, before the western half of the state was opened to white settlement.

Opportunities in mining attracted the metal miners of England and northern Europe to the Black Hills. They brought strong backs, a will to work, and expertise to the mines. Opportunities for graduates of European technical schools attracted mining engineers, metallurgists, and geologists to the northern Black Hills. Many of the professionals who staffed the state School of Mines at Rapid City received their scientific and engineering training in central Europe.

The demand for foodstuffs, especially fresh meat, led to establishment of a cattle and sheep industry in the western rangeland. Instead of stopping all the big Texas cattle drives at the railroad shipping points in Kansas, some pioneer trail drivers continued to western South Dakota to furnish food for the miners and later for the reservation Indians after the buffalo were gone. Capital from England and Scotland encouraged rapid stocking of large cattle and sheep ranches in western South Dakota and adjacent states.

The early years of the 20th century brought the opening of large areas of Indian land to white settlement; northern Europeans came by the trainload to take up homesteads. The small size of the allotments, poor soil, more rugged terrain, and marginal rainfall all conspired against the western homesteaders. Much of their land reverted to cattle and sheep raising as the mainstay of western agriculture.

Rocks

The rocks of South Dakota are highly varied. In some places, you see towering cliffs, in others a lone butte on the skyline. In many places, the rocks appear in the slopes of a steep valley, more commonly only in a roadcut or borrow pit. Most of the rocks are concealed beneath the prairie sod. Knowledge of the rocks beneath the surface is important for the development of groundwater supplies in this dry region and for exploration of coal, oil, or minerals.

Human beings have long regarded the earth's major features to be as everlasting as the hills. Now it is clear that the continents move as parts of large segments of the earth's outer rind called plates. In some places, plates pull away from each other; in other places they collide; in still others they slide past each other. Most geological activity seems to happen at or near plate boundaries. South Dakota is near the middle of the great North American plate, seemingly little affected by the great global forces that move the plates. The Black Hills may have risen as a result of plate movement, but it is not clear how.

The earth probably formed about 4.6 billion years ago. Until recently, the oldest rock that geologists had dated was about 3.8 billion years old, but one from western Australia contains crystals of zircon 4.3 billion years old.

Christians divide the history of man into two very unequal periods of time: the pre-Christian epoch that begins with the advent of our species, and the Christian epoch that begins with the birth of Christ. Geologists somewhat similarly divide geologic history into Precambrian time, which begins with the oldest rocks, about 4 billion years ago, and the time in which animals have lived on earth, which begins with the Cambrian period, about 570 million years ago.

The oldest rocks in South Dakota are granites and contorted metamorphic rocks that formed during Precambrian time, most of them more than 2 billion years ago. You can see them in the core of the Black Hills and in two small areas in eastern South Dakota. Elsewhere, those oldest rocks lie buried under a sequence of sedimentary rocks deposited over the last 520 million years. Except in the Black Hills, most of the layers still lie almost as flat as on the day they were laid down. Almost, but not quite. They actually dip at

Era	Period	Epoch	Principal Events in South Dakota
Cenozoic Age of Mammals	Quaternary	Holocene	Widespread erosion
		— 0.01 my —	
		Pleistocene	Glaciation in eastern South Dakota, erosion in western area
		— 2 my —	
	Tertiary	Pliocene	More erosion than deposition
		— 5 my —	
		Miocene	Continued deposition of light-colored clays and sands over much of the area covered by Oligocene deposition, but in less volume and not so widespread
		— 24 my —	
		Oligocene	Deposition of a thick apron of light-colored sands and clays over western South Dakota; erosion in extreme eastern part of the state
		— 38 my —	
		Eocene	Continued erosion of Black Hills; most debris carried eastward beyond limits of South Dakota
		— 55 my —	
		Paleocene	Uplift of the Black Hills, with igneous activity in northern area
		— 63 my —	
Mesozoic Age of Reptiles	Cretaceous	— 138 my —	Thick deposits of black mud over most of South Dakota
	Jurassic	— 205 my —	Erosion in lower and middle Jurassic, deposition in upper Jurassic
	Triassic	— 240 my —	Deposition of redbeds at least in western third of state
Paleozoic Age of Fishes	Permian	— 290 my —	Shallow-water marine deposition in western South Dakota; probable erosion in eastern section
	Pennsylvanian	— 330 my —	Marine deposition of sandstones, carbonates, and evaporites
	Mississippian	— 365 my —	Erosion except in north-central part of state; limestone in lower Mississippian
	Devonian	— 410 my —	Deposition of carbonates in Williston basin extended to central South Dakota
	Silurian	— 435 my —	Deposition as in Devonian
	Ordovician	— 500 my —	Limestone deposited over much of state in upper Ordovician period; lower Ordovician deposition of sand and shale continued from late Cambrian time; erosion in lower and middle Cambrian before the sea reached as far east as South Dakota
	Cambrian	— 570 my —	
Precambrian	Proterozoic	— 1,715 my —	Long period of erosion
		— 2,500 my —	Emplacement of Harney Peak granite; deposition of marine sands and clays
	Archean		Intrusion of Little Elk and Bear Mountain granites; deposition of shales interrupted by periods of mountain building and volcanic activity

my = million years

Geologic time scale and principal events in South Dakota.

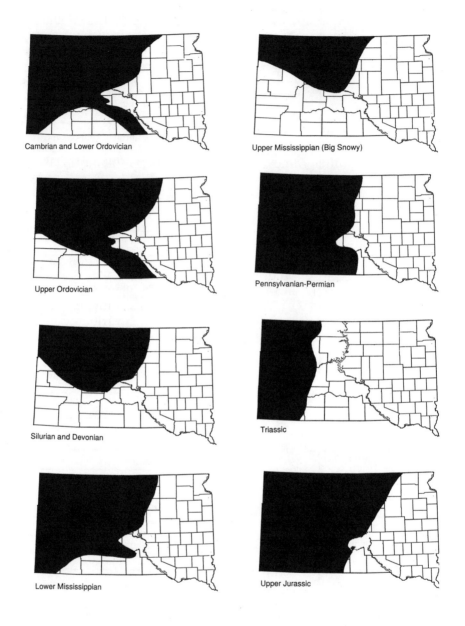

Cambrian and Lower Ordovician

Upper Mississippian (Big Snowy)

Upper Ordovician

Pennsylvanian-Permian

Silurian and Devonian

Triassic

Lower Mississippian

Upper Jurassic

Distribution of Paleozoic and Mesozoic sedimentary rocks in South Dakota.

angles of about 30 feet to the mile.

Rocks of all kinds and of all ages are composed of mineral grains, either one kind or several. Some rocks are very hard; others, including many in South Dakota, are soft. Either way, they are rocks in the eyes of geologists, who generally divide them into three main groups: igneous, sedimentary, and metamorphic.

Igneous rocks crystallize from molten magma, either at the surface or at depth. Volcanic rocks generally consist mostly of mineral grains too small to see without a strong magnifier; rocks that crystallize at depth consist mainly of grains large enough to distinguish without a magnifier. Almost all the igneous rocks in South Dakota crystallized at depth and are fairly coarsely granular. A few of the igneous rocks in South Dakota contain large crystals set in a matrix of much smaller grains. That is called a porphyritic texture. Such rocks abound in the younger bodies of igneous rocks in the northern Black Hills.

Sedimentary rocks consist of debris that accumulates on the bottom of a sea or lake, or on dry land. The debris may have originated from weathering and erosion of older rocks, or it may consist of material that formed within the sedimentary environment, such as seashells. Streams, wind, waves, and glaciers all contributed material to the sedimentary deposits of South Dakota. Most sedimentary

Porphyry. —Phil Bjork, S.D. Museum of Geology

rocks are soft when freshly deposited, then solidify in their own good time. Most sedimentary rocks are conspicuously layered, which makes them easy to recognize, even at a distance.

Shale is the most abundant sedimentary rock type in South Dakota. It consists mostly of clay, perhaps mixed with small amounts of silt or sand. Shale tends to split into thin flakes. Most of the shale in South Dakota was deposited in shallow seawater.

Bentonite is an uncommon sedimentary rock that forms from volcanic ash that falls into seawater. In South Dakota, it generally appears as layers of light yellow to greenish clay sandwiched within dark shales.

Sandstone, another abundant sedimentary rock in South Dakota, consists mainly of sand, exactly as the name suggests. If the sand is very tightly cemented into very hard rock that breaks across the sand grains instead of around them, you can call it quartzite. Siltstone is like sandstone except that it consists of much smaller grains. You can tell the difference by dragging your fingernail across the rock: If it feels sandy, it is sandstone; if it feels more like a lady's fine emery board, it is siltstone.

Coal forms from peat, a mass of more or less decomposed plant material laid down in some sort of swamp or marsh, most typically along a coastline. When peat is buried, the weight of the overlying sediments compresses it into coal. The kind of coal depends mainly upon how deeply it was buried. Most of the coal in South Dakota is lignite, generally rather poor quality stuff.

Some rocks form as minerals precipitate directly from water. The commonest of these in South Dakota is gypsum, calcium sulfate.

Limestone consists mainly of the mineral calcite, calcium carbonate. Some kinds of limestone also precipitate directly from a water solution, but most kinds consist largely of fragments of seashells cemented together. Dolomite is like limestone, except that it contains a large proportion of magnesium instead of calcium. Most of the limestone in the Black Hills is more or less dolomitic. Limestone typically forms on shallow seafloors or in shallow lakes.

Metamorphic rocks form when sedimentary or igneous rocks recrystallize at high temperature, generally under high pressure. The original minerals may simply recrystallize into larger grains, or they may transform into entirely new metamorphic minerals. And their minerals generally align parallel to each other, as though they were in military formation. Flakes of mica, for example, are parallel to each other like the pages in a book, and needles of black hornblende are parallel like a fistful of pencils. That alignment gives metamorphic rocks a directional grain, like the grain of wood. The core of the Black Hills is the only part of South Dakota where you can see metamorphic rocks. Schists are the most abundant; most of those in the Black Hills are gray, and they contain enough mica to give the rock a flaky quality. They tend to split into slabs. Some of the schists contain beautiful crystals of red garnet, or little crosses of staurolite. Other metamorphic rocks include marble, which is recrystallized limestone or dolomite, and gneiss, which looks like granite except in having that streaky metamorphic grain. The Black Hills also contain dark amphibolite gneisses, which are recrystallized basalt.

Rocks to Soil to Sediment

The processes of weathering convert rocks to soil, and the processes of erosion carve the landscape as they carry the soil away.

Water attacks the silicate minerals in igneous and metamorphic rocks, converting them into clay. The minerals swell as they become clay. That breaks the rock apart just as a wall would fall apart if most of its bricks were to swell up. Only quartz is immune to the attack of water; it survives as sand grains.

Meanwhile, atmospheric oxygen attacks the iron in the rock, especially the iron sulfide mineral pyrite, converting it into iron oxide, or rust. Iron oxide is an excellent pigment familiar in red barn paint and red primers. It stains rocks and soils red or brown.

Limestone, dolomite, and marble simply dissolve in slightly acidic rainwater. Some formations dissolve at the surface, others within to form caverns. All caverns eventually collapse, and many break through to the surface to become sinkholes.

Running water is the main agent of erosion. Even small streams and rivulets carry loose material in suspension and roll it along their course. Rivers carry huge loads. Floods can perform spectacular feats in moving debris of all sizes. Wind excavates shallow depressions in the plains and eventually dumps the eroded sediment in deposits of dust called loess, which makes excellent soil.

Water

Water is a precious commodity in South Dakota, especially west of the Missouri River. Rainfall ranges from about 30 inches per year in the southeast corner to less than 16 inches in the southwest. Locally, on the Black Hills uplift, rainfall exceeds 20 inches. Except for the Missouri River, even the major streams may cease to flow in late summer. Countless stock dams retain snowmelt and water from summer showers. In a few areas, rural water systems deliver water from the Missouri River reservoirs.

Most towns, farms, and ranches depend upon wells for their domestic water supply. In the glaciated area east of the Missouri River, groundwater is abundant in glacial outwash gravels and along old meltwater stream channels. But glacial till retains its water, yielding it only sparingly to wells. In the vast areas west of the Missouri River where Pierre shale forms the bedrock, water is scarce. The shale provides virtually none, and many people must rely on deep wells. It was a real boon to settlement back in the 1880s, when someone discovered artesian water in sandstones beneath the shales at depths ranging from a few hundred to more than 2,000 feet.

The simplest artesian system consists of a widespread layer of porous rock such as sandstone confined beneath a layer of impermeable rock such as shale. If the porous aquifer rock crops out at the surface at a high elevation, as it does in the Black Hills, water may fill it and then move down, away from the intake area. Then a well drilled through the overlying impermeable shale will find the sandstone aquifer filled with water at a pressure that corresponds to the elevation of the area where it filled. The water will rise in the well bore and may flow to the surface, unpumped. That is an artesian well, named for the province of Artois in northern France, where an artesian aquifer was discovered and developed in the 12th century.

If it were not for the friction of water flowing through the minute openings in the aquifer, water in drilled wells would rise to the same elevation as the water at the intake area, as it does in the spout of a teakettle. The surface defined by the height to which the water actually rises is called the pressure, or potentiometric, surface. It slopes gently downward away from the intake area. Where that pressure surface lies above the contours of the land, a well will flow; where it is below land surface, the water will rise to some point below the surface and must be pumped the rest of the way. By comparing care-

An idealized artesian system.

intake or
recharge area

artesian well

level to which artesian water will rise

ground surface

impervious layer (shale)

impervious layer (shale)

fully prepared contour maps drawn on the pressure surface with those of the land surface, it is possible to predict ahead of drilling whether a well will flow and the approximate water pressure.

Several famous artesian aquifers lie beneath the plains of South Dakota. The most widespread is the Dakota sandstone aquifer. Near the Black Hills, the sandstones in the upper part of the Minnelusa formation are an important source of artesian water. The Pahasapa, or Madison, limestone is an important source of water in the western part of the state. The intake or recharge areas for each of these aquifers is in the Black Hills at elevations higher than most of the state. Broad outcrop areas of each permit water to fill them. Huge flows of water enter the Madison limestone where streams sink into caverns. Some of that water returns to the surface through large springs, mainly in the Red Valley.

The quality of artesian water varies greatly, depending upon the quality of the water that enters the aquifer and what happens to it afterward. Artesian water in eastern South Dakota commonly contains enough calcium to qualify as hard water. One sand layer may contain hard water, another soft water. Ingenious well drillers place one pipe within another so that one type of water will come up the inner pipe, another through the annular space between the two pipes.

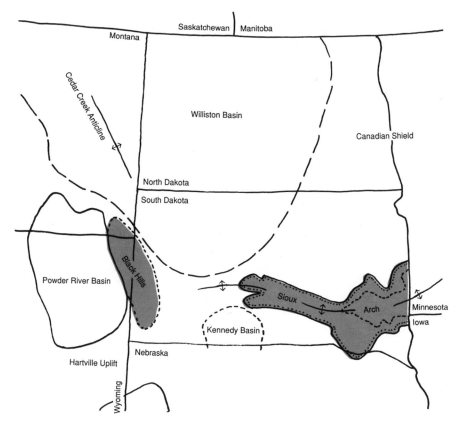

Major structural features of South Dakota.

Rocks and Landscapes East of the Missouri River

Ancient Rocks

Extremely ancient rocks like those that lie buried under most of South Dakota are at the surface in much of central Canada, an area known among geologists as the Canadian Shield. It extends south through the northern half of Minnesota and into very small areas of northeastern South Dakota. Much of the sand in the sediments in eastern South Dakota probably came from the ancient rocks of the shield.

A very broad and very gentle arch that extends from southeastern Minnesota to central South Dakota apparently reflects an elongate ridge of very ancient quartzite. It is the Sioux quartzite, a remnant of a blanket of sand that was laid down across a large region about 1.7 billion years ago. By about 600 million years ago, the beginning of Cambrian time, it was eroded to a long ridge that stood as much as several hundred feet above the complex of older rocks on which it was deposited. That was when the sedimentary formations that cover the older rocks began to accumulate. Now, that buried ridge of Sioux quartzite shows through the blankets of younger sedimentary formations, like a dog hiding under the covers, to make the Sioux arch. In some places in eastern South Dakota, generally between Sioux Falls and Mitchell, the Sioux quartzite is exposed, uncovered, at the surface.

There you can see that it is a pink sandstone, so tightly cemented that geologists call it quartzite. It owes its color to a film of iron oxide on each grain of sand. Ripple marks, mud cracks, and raindrop impressions show that it was deposited in shallow water and on dryland.

The Glacial Period

Pleistocene time began about 2 million years ago. World climate then began a series of fluctuations, periods of glaciation alternating with interglacial periods when the glaciers melted and the climate warmed. At least four major periods of glaciation are documented—with many minor fluctuations. We do not know whether the present

11

warm climate is another interglacial period, with another ice age in prospect, or whether the last ice age is behind us.

When snow and ice collect faster than they can melt, an ice cap starts to form. In the Hudson Bay area, the ice cap built to an estimated thickness of about 10,000 feet, as judged from ice caps on Greenland and Antarctica. That enormous weight makes the ice flow. Such continental ice glaciated most of the northeastern United States and adjacent parts of Canada and covered eastern South Dakota several times.

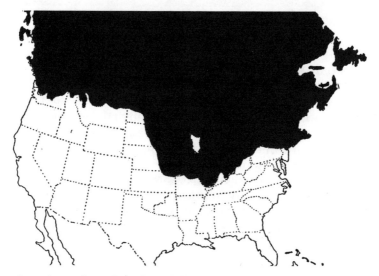

Areas in southern Canada and the northern United States covered by Pleistocene glaciers.

As the ice sheet moved, much of the rock debris in its path froze into the lower layers of ice. When the glacier reached its southern limits, it carried a huge load of debris that it dropped as the ice finally melted. Where the ice sheet was very thick and heavy, the glacier scoured and smoothed off the terrain. Where the ice was thin, it overrode obstacles rather than planing them off.

Glacial Deposits

The material transported by the glaciers consists of ground-up rock and boulders. Farmers call it boulder clay; geologists call it glacial till. Any deposit of glacial till is called a moraine. Where debris was dropped from the bottom of a moving glacier, the continued movement of the ice smoothed off the surface, leaving a gently un-

Pothole lakes, McCook County. —U.S. Fish and Wildlife Service

dulating deposit of debris—a ground moraine. Where stagnant ice melted, a stagnation moraine formed. Its surface is hummocky and irregular, partly because moving ice never planed it smooth. Random masses of stagnant ice, buried in the ground moraine, melted last, leaving depressions that became pothole lakes.

Where an ice front remained more or less stationary for a long time, great morainal ridges of glacial till piled up, exactly recording the former outline of the glacier. Those moraines are the most prominent glacial features and the most informative.

When a glacier melts, tremendous volumes of water flow as countless streams on and beneath the ice. Water pouring off the ice deposits

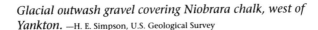

Glacial outwash gravel covering Niobrara chalk, west of Yankton. —H. E. Simpson, U.S. Geological Survey

a broad outwash plain of sand and gravel that stretches beyond the moraine. Outwash gravels are commonly well sorted by size. Coarsest materials move the least, while fine clays flush away in suspension. This winnowing action produces commercially valuable accumulations of boulders, gravels, and sand.

When outwash plains dry out, turbulent winds along the ice front pick up the fine material, creating huge dust storms. Away from the front, this dust settles to form deposits called loess. Because loess consists mostly of sharp particles of pulverized rock, the grains interlock. So the deposits are very stable and commonly stand in bluffs tens of feet high.

Eskers are long, sinuous ridges of gravel deposited by streams flowing beneath the ice or in crevasses within the glacier. Kames are isolated mounds of gravel formed where streams of meltwater poured out of the end of the glacier or into low places on its surface.

Glacial History

The study of glaciers and glacial deposits developed first in Europe. In this country, the upper Mississippi Valley area is the most completely studied glacial sequence. Workers try to correlate deposits in other areas with those in this classic area where a detailed series of glacial and interglacial events has been developed.

In recent years, it has become increasingly clear that this scheme of events during the glacial epoch is probably not correct. It now

Roadcut in an old kame terrace north of Yankton. —H. E. Simpson, U.S. Geological Survey

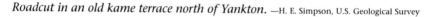

seems that many more than four ice ages came and went during Pleistocene time, but no one is quite sure how many, or when. So geologists find themselves in the awkward position of knowing that their old way of doing things is probably wrong, but without having a new way. So we will use this old scheme here because it is still all we have.

Wisconsin Ice Age
Mankato substage (ended 10,000 to 15,000 years ago)
   Cary-Mankato interglacial
Cary substage
   Tazewell-Cary interglacial
Tazewell substage
   Iowan-Tazewell interglacial
Iowan substage (began 60,000 years ago)
   Sangamon interglacial

Illinoisan Ice Age
Yarmouth interglacial

Kansan Ice Age
Aftonian interglacial

Nebraskan Ice Age

The various glacial advances covered different areas. They are named for areas where later advances did not destroy or bury them. Nebraskan ice extended into eastern Nebraska as well as into northern Missouri. Kansan ice overran much of the same area and extended into Kansas. It overrode and in many instances partially destroyed the deposits left by the earlier Nebraskan ice sheet. The third glacial advance, the Illinoisan, extended nearly to the southern tip of Illinois and into eastern and central South Dakota. Features left by the ice of the final stage of the Wisconsin ice age are best known because they are the least disturbed or eroded. That final stage started about 60,000 years ago and ended just 10,000 to 15,000 years ago. Geologists divide it into four substages.

The Glacial Sequence in Eastern South Dakota

During the Wisconsin ice age, the great continental glacier advanced several times into eastern South Dakota. We identify each advance by the sequence in which one layer of glacial debris covers another. The radiocarbon method can date specimens of wood or charcoal from deposits less than about 50,000 years old. South Dakota geologists now recognize the following drift sheets, but their correlation with the Mississippi Valley section is not entirely clear.

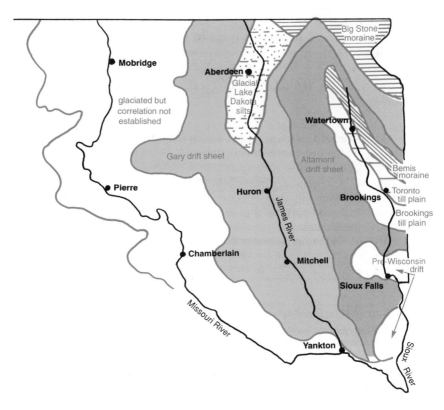

Major glacial landforms in eastern South Dakota.

Isolated glacial boulders occur in an irregular band up to 50 miles wide west of the Missouri River. They testify to an advance of ice westward before the formation of the Missouri Valley trench. Only boulders remain: The till was eroded away. Such boulders not related to the bedrock upon which they remain are called glacial erratics.

Glaciers Move the Rivers

Before Pleistocene time, the major streams of western South Dakota flowed east, as they now do, but on an old surface 100 to 200 feet above the present upland surface. We know this because we find old stream gravels on the flat tops of many buttes. The gravels contain fossils, particularly horse teeth that date from the middle of Pleistocene time. In eastern South Dakota, that old surface lies buried beneath glacial deposits.

Sequence of glacial events in eastern South Dakota

Big Stone stage (=Wisconsin?): This youngest stage advanced only into the Minnesota River lowland in the extreme northeastern corner of the state.

Gary stage: This sheet split at the top of the high-standing Prairie Hills, or the Coteau des Prairies. The east lobe, called the Des Moines lobe, lapped against the east side of the Coteau, but did not over-run it. The west lobe followed down the James River Valley to the southern boundary of the state near Yankton, but it did not encroach upon the highland of the Coteau.

Altamont stage: Ice did not advance onto the eastern side of the Prairie Hills, but left a persistent ridge of morainal material along the eastern edge. The lobe that went down the James River lowland encroached nearly halfway onto the western side of the Prairie Hills.

Bemis stage: We have identified drift of this age only on the eastern side of the Prairie Hills, where it left a high belt of moraine on the eastern crest.

Toronto till plain: The oldest Wisconsin drift positively identified in eastern South Dakota occupies a narrow strip along the central part of the Prairie Hills. It escaped burial by later ice from either side of the Prairie Hills.

Brookings till plain: This belt of deeply weathered ground moraine extending from northwest to southeast through Brookings also escaped burial from either of the Wisconsin lobes. It is probably Illinoisan in age.

Older or uncorrelated drift: Extending beyond the southern and western limits of the Gary ice sheet in the James River lowland and extending to and beyond the Missouri River is a large expanse of till. Its age is uncertain, but probably middle Wisconsin.

Oldest drift: Two areas in southeastern South Dakota are overlain by deeply weathered drift of Illinoisan or even pre-Illinoisan age. One is at Sioux Falls, the other in the Newton Hills along the Big Sioux River in the extreme southeastern corner of the state.

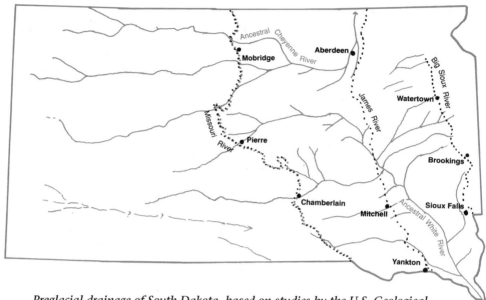

Preglacial drainage of South Dakota, based on studies by the U.S. Geological Survey and the South Dakota Geological Survey.

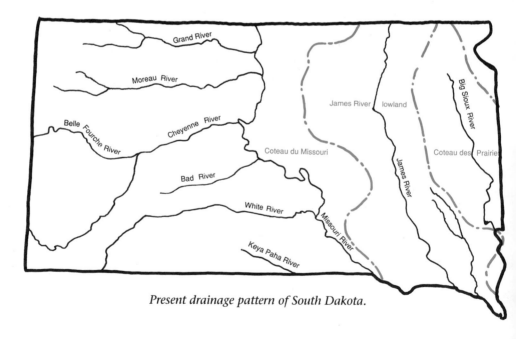

Present drainage pattern of South Dakota.

If we could strip off that glacial cover east of the Missouri River, we would almost certainly see that those streams once continued eastward across the state. Rivers in the northern portion of the state—the Grand, the Moreau, and probably the Cheyenne—swung northeastward and ultimately joined a major river that drained north into Hudson Bay. Streams in the southern portion of the state—the Bad, the White, and the Niobrara—swung southeastward, and entered an ancestral Missouri River that discharged into the Mississippi River. The continental divide between waters flowing north and those flowing south was near Redfield in Spink County. The old stream pattern is generally preserved beneath the glacial drift, but was destroyed where an earlier glacier scoured out the James River basin.

During middle Pleistocene time, the Illinoisan ice sheet covered all of eastern South Dakota and effectively dammed the streams that flowed east. Their waters flowed around the southwestern edge of the ice sheet and joined the ancestral Missouri River near Yankton. These escaping waters, supplemented by water from melting glaciers, rapidly scoured this new channel of the Missouri River to a depth of 300 to 700 feet below the old upland surface. Thus, when the Illinoisan glacier melted, the western streams continued to feed into the new Missouri River because they could not flow uphill into their former channels to the east.

When the streams on the eastern side of the Black Hills began discharging into the newly formed trench several hundred feet below their former channels, their velocity increased. They started eroding downward and headward through the soft, easily erodible Pierre shale. The Cheyenne River, in particular, became very active. Meanwhile, the Belle Fourche River worked its way headward around the northern end of the Black Hills. There it captured the headwaters of the Little Missouri just west of the South Dakota line at Alzada, Montana. A short, southern fork began working headward in a southwesterly direction and successively captured Elk, Box Elder, Rapid, Spring, Battle, and Grace Coolidge Creeks on the eastern side of the Black Hills. It then flowed around the southern end of the Black Hills to drain a small area in eastern Wyoming.

Major Topographic Features of the Glaciated Area

A visitor's first impression of eastern South Dakota is that of a gently rolling to flat plain covered with glacial till or other glacial debris. A glance at a high-altitude aerial photograph, or even a topographic map, shows half a dozen very large features. These divide the region into distinct geographic areas.

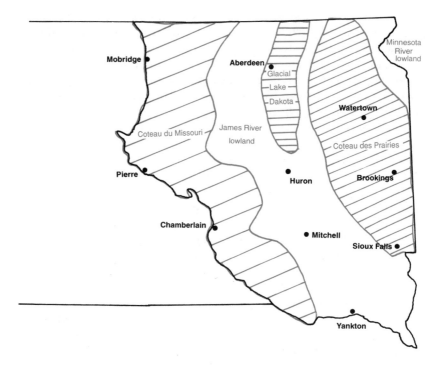

Major topographic features of the glaciated area.

In the northeastern corner is a small area called the Minnesota River lowland, which includes the lowest point in the state. The lowland also includes a continental divide separating waters draining to Hudson Bay from those heading for the Gulf of Mexico.

A much larger area got its name from the early trappers, who called it the Coteau des Prairies, or the Prairie Hills. It is a broad trough whose margins stand 500 or more feet above the general prairie. It crosses the state from north to south and is as much as 60 miles wide.

The James River lowland stretches across the middle of eastern South Dakota from north to south. An early ice advance scoured it out of the soft shale bedrock. It is bounded rather sharply on the east by the Prairie Hills, less sharply on the west by the Missouri Hills, which the early trappers called the Coteau du Missouri. The last glacier, unable to override the highlands on either side, pushed southward through the lowland. As its front finally melted back, it did so in stages, leaving a small moraine to mark the places where it briefly stabilized. These small moraines cross the entire length of the James River lowland in a series of more or less parallel ridges. Meanwhile, the James River carried the meltwaters toward the Gulf of Mexico.

At one stage, the ice front stabilized near Redfield long enough to form a fairly substantial moraine. It was a natural dam that impounded the meltwater in a vast temporary lake that geologists call Glacial Lake Dakota. In its prime, it was more than 100 miles long and as much as 30 miles wide. When the impounded water finally breached the morainal dam, a great flood rushed down the James Valley, eroding the sharply defined inner valley, a trench some 60 to 100 feet deep and from 0.5 to 1 mile wide.

When the ice had finally melted, the only water to feed the river came from rainfall. Now we have a puny stream, wildly meandering, lost in the broad floor of the James River Valley. The present James River system is about 250 miles long, but if all the meanders were stretched out, the true length of the stream would be more than 700 miles. The total drop in elevation is only 100 feet, so the stream is sluggish. It takes water three weeks to cross South Dakota.

Since the ice retreated a mere 10,000 years ago, the James River has not yet succeeded in draining its entire basin. In many areas, rainfall soaks into the ground or runs off into small lakes between morainal ridges. It will take several thousand more years to establish the full stream pattern within the James River basin.

Between the James River lowland and the deep valley of the Missouri River is a sinuous strip of high ground about 25 to 35 miles wide, the Coteau du Missouri. It is a strip of the Great Plains separated from the plains farther west by the Missouri River. The crest of the strip is a series of bold glacial moraines along its eastern edge. Segments of this ridge have local names such as the Bowdle and Lebanon Hills. Farther south are Ree Heights, the Wessington Hills, and finally the Bijou Hills, 35 miles north of the southern boundary of the state. West of the crest are rolling plains in which a thin veneer of glacial debris covers the Pierre shale. Steep streams deeply dissect the ridge along its western margin, cutting through the glacial deposits and into the Pierre shale. Bedrock under the Missouri Hills stands several hundred feet above that beneath the James River basin. The cover of glacial till is generally less than 100 feet thick.

Road Logs East of the Missouri River

I-29
Sioux City, Iowa–North Dakota Line
267 miles

Glacial ice covered all of South Dakota east of the Missouri River several times. The last ice sheet melted from the northeastern corner of the state less than 10,000 years ago. Glacial debris, from a few tens of feet to several hundred feet thick, conceals the bedrock in all but a few places.

Except for a few miles at either end, I-29 traverses the length of the Prairie Hills, a rolling plateau that rises 300 to 600 feet above the Minnesota River lowland to the east and the James River lowland to the west. Its dimensions are roughly 200 miles from north to south and up to 70 miles from east to west. In shape, it is a broad trough, highest around the margins, lowest in its central drainage area along the Big Sioux River. I-29 roughly follows that river, which was a major drainageway when the ice sheet was melting.

Geologic features here are interesting, but not spectacular. Most of the surface is a gently rolling plain underlain by several hundred feet of glacial till left as the ice melted. Long moraines of glacial till, deposited when the ice front remained stationary for long periods, stand above the general surface. Countless stream valleys, cut by glacial meltwater flowing toward the major streams, entrench the general plain. The valleys are too large for the streams that now flow through them, and they are deeply filled with glacial outwash gravel. Small lakes abound. Most are potholes.

Between Sioux City and Sioux Falls

The highway crosses a broad alluvial plain between the Missouri River on the west and the Big Sioux River on the east. The old valley, which predates the glacial period, ranges from 6 to 9 miles wide. It provides some of the most fertile farmland in the state. Scattered crescent-shaped lakes and sloughs are remnants of meanders in earlier channels of the Missouri River.

The wind blew sandy stream deposits from the river bottom into an area of sand dunes, the Dakota Dunes, now partially stabilized. They have become a favorite local recreation area.

Just south of Junction City, I-29 leaves the valley bottom and climbs onto an older upland surface that exhibits typical rolling glacial topography. The landscape is a flat to gently rolling ground moraine from Junction City to the southern edge of the Sioux Falls metropolitan area. East of I-29, the Newton Hills lie along the skyline for nearly 30 miles. The Big Sioux River borders the Newton Hills on the east; Brule Creek borders them on the west. This high and rough area is an isolated southern extension of the Prairie Hills. Bedrock under

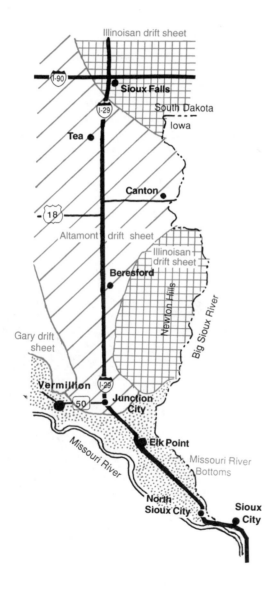

I-29, Sioux City, Iowa, to Sioux Falls, South Dakota (83 miles).

the ridge is Niobrara chalk, a limestone deposited in the shallow inland sea that flooded this area during late Cretaceous time.

An earlier glacier overran the Newton Hills, but did not have enough weight or energy to plane them off. It left a thin mantle of very old and deeply weathered glacial till over the surface. This, in turn, has been partially concealed by several feet of windblown fine silt, or loess. From the north end of the Newton Hills to the vicinity of Sioux Falls, a lowland about 15 miles wide separates this southern outlier from the main Prairie Hills.

The conspicuous ridge that the highway crosses about 6 miles north of Beresford is a glacial moraine. I-29 enters the Prairie Hills just south of Sioux Falls, but the climb is subtle because the route follows the valley of the Big Sioux River.

A small meteorite that weighed 1.6 ounces fell in Centerville on February 29, 1956. It penetrated the sheet aluminum roof of a barn and lodged in a corn planter. Both the stony meteorite and a small section of the barn roof are on display at the Museum of Geology at the School of Mines in Rapid City.

At Sioux Falls, the glacial drift is thin, missing in places. A ridge of very resistant Sioux quartzite, a Precambrian sedimentary formation, creates a series of falls and rapids along the Big Sioux River.

The falls at Sioux Falls. —D. G. Jorgensen, U.S. Geological Survey

Between Sioux Falls and Watertown

I-29 generally parallels the Big Sioux River. For 25 miles north from Sioux Falls, the road crosses some of the oldest glacial drift in the state. It was probably laid down during the Illinoisan glaciation. Scattered stream valleys, much choked with gravel, cut the gently undulating surface. They carried huge volumes of meltwater away from the ice fronts as the glaciers rapidly melted at the end of the ice age.

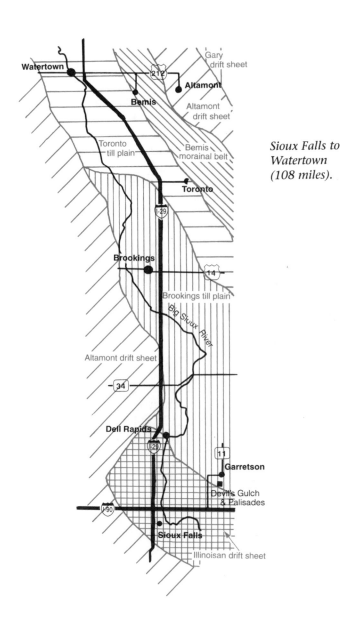

Sioux Falls to Watertown (108 miles).

Dell Rapids

Three miles east of I-29, SD 115 crosses the Big Sioux River. Just south of Dell Rapids, the river has carved a picturesque gorge through a ridge of Sioux quartzite, creating the Dells of the Big Sioux River. In places, the gorge is 80 feet deep. A huge flood during late glacial time plucked blocks of quartzite out of the ridge. A good secondary road follows the rim of the Dells gorge, which extends along the river for about a mile and a half.

In a stretch of 10 to 15 miles north and south of the junction with SD 34 at Exit 109, the highway skirts the western edge of an area of somewhat younger glacial drift. Watch for the very distinct ridges of the terminal moraines.

Flandreau Meteorite

A very large meteorite was found near Flandreau in 1981. No one knows when it fell, presumably many years ago, long before anyone in South Dakota was keeping written records. It is a stony meteorite that consists entirely of the minerals olivine and bronzite, which are common in black igneous rocks.

Like most stony meteorites, this one consists mostly of little spheres about the size of bird shot called chondrules. They appear to have condensed from a hot cloud in space in much the same way that droplets of water condense in a summer storm cloud. No earthly rocks contain chondrules, only stony meteorites; that is how geologists recognize them.

Between Exits 125 and 150, the highway crosses the Brookings till plain, an area of glacial deposits that were laid down before the Wisconsin glaciation, during some previous ice age. Few distinctive glacial features survive, except for old meltwater channels now occupied by small streams.

The Brookings plain is the only part of the Prairie Hills that was not covered at one time or another by advances of late Wisconsin ice. Brookings is on an island of old ground moraine completely surrounded by younger meltwater channels deeply filled with gravel.

The section of highway between Exit 150 and Watertown trends northwest, across the Toronto glacial plain. The thickness of the glacial debris ranges from 400 to 600 feet. The succession of layers revealed in test holes shows that several successive advances and retreats of glacial ice laid down this thick blanket of debris. Outwash gravels accumulated on the surface between the times when ice covered it. The low ridges of the Bemis belt of terminal moraines parallels this section a few miles to the northeast. They mark the limits of the Bemis ice advance in South Dakota.

Between Watertown and North Dakota Line

Between Watertown and Exit 201, the highway crosses the part of the Toronto till plain that was plastered with ground moraine early in the Wisconsin ice age. The only distinctive features are a few small pothole lakes formed where blocks of ice trapped in the drift or in outwash gravels melted.

Between Exits 201 and 207, the highway cuts diagonally across the Bemis terminal moraine belt. The town of Summit occupies a high point on the eastern edge of the Prairie Hills. The Altamont moraine on which it stands records an attempt of a glacier pushing

Watertown to North Dakota line (76 miles).

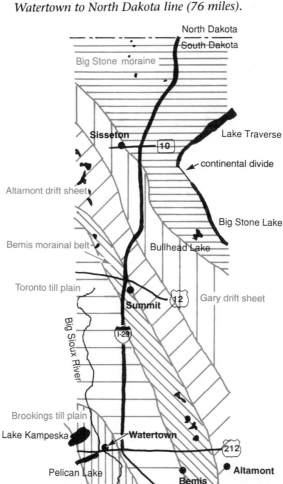

Glacial boulders weathering out of a moraine in Roberts County.

Pothole lake. —S.D. Geological Survey

down the Minnesota River lowland to overrun the eastern edge of the Prairie Hills.

North of Summit, the highway diagonally crosses the eastern face of the Prairie Hills escarpment and enters the Minnesota River lowland. The change in elevation is about 750 feet in 10 miles.

Between the area north of SD 10, near Sisseton and the North Dakota line, the surface is a gently undulating plain, generally less than 1,200 feet above sea level, dotted with potholes and sloughs. Near the intersection with SD 10, the highway crosses a subtle continental divide. Drainage south of SD 10 enters the Minnesota River and ends up in the Gulf of Mexico, whereas that to the north drains through the Red River of the North, ultimately into Hudson Bay.

Glacial sediments in this corner of the Minnesota River lowland record the last invasion of glacial ice into South Dakota. The Big Stone glacier advanced only a few miles into the state before it melted, less than 10,000 years ago.

I-90
Minnesota Line–Missouri River
152 miles

Between the Minnesota state line and the Missouri River, I-90 crosses three geographic provinces. The eastern province is the high Prairie Hills from the state line to the east fork of the Vermillion River between Humboldt and Montrose. The middle province is the James River lowland from the east fork of a conspicuous ridge between Kimball and Pukwana. The western province is the slightly higher and gently rolling Missouri Hills from the vicinity of Pukwana to the Missouri River. Some geographers include the deep Missouri River trench as a fourth province.

Glacial Debris
Thick sheets of glacial ice covered this entire area several times. The ice moved down from North Dakota and Minnesota, away from the much thicker accumulations farther north. The last ice melted from eastern South Dakota about 10,000 years ago. The melting ice dropped the rock and clay debris from within it, leaving a heterogeneous blanket of glacial till and outwash sands and gravels. Thicknesses of the debris range from a few feet to a few hundred feet. Collectively it is all glacial drift, if you do not care to distinguish

between till and outwash. Till is material dropped directly from the ice; outwash was carried in glacial meltwater.

Any deposit of till is a moraine. The principal kinds of moraines in this area are ground moraine, stagnation moraine, and end moraine.

Ground moraine is deposited from moving ice and includes material bulldozed ahead of the advancing ice front. The moving ice rudely leveled the surface, leaving a gently rolling topography. Stagnation moraine is dumped from ice that has stopped moving and is melting in place. It makes rough terrain, full of sloughs and potholes. End moraine consists of ridges deposited along the front of the glacier. These ridges record times when the rate of ice advance was in balance with the rate of melting, so the ice front remained nearly stationary. These long ridges of glacial till may be several hundreds of feet high and hundreds to thousands of feet wide.

Glacial outwash typically forms broad plains where great torrents of meltwater, issuing from under the glacier and along the ice front, dropped the heavier part of their loads of debris. Water carried away most of the finer material, leaving great outwash sheets or valleys full of coarse sand and gravel.

Loess is windblown dust. When outwash plains dried as the meltwater diminished, strong winds blowing off the ice whipped up great clouds of dust from the surfaces of the plains and dropped it miles away. It settled as a blanket of silt, from a few inches to many feet thick, conforming to the topography. The surface is absolutely free of rocks and makes ideal farmland.

Long periods in which the climate was as warm as or warmer than today separated the major advances of the ice. Along I-90 we enter the state on till of Illinoisan age for a few miles; the rest of the way to the Missouri River is on glacial deposits of Wisconsin age. Several minor advances and retreats of the ice affected South Dakota in Wisconsin time. We cross a narrow band of slightly older Altamont moraine west of Sioux Falls. West of White Lake, we will drive over an earlier Wisconsin till whose exact age is uncertain.

Sioux Quartzite

Bedrock under the glacial drift between the state line and Alexandria is the extremely ancient Sioux quartzite. Its outcrops make spectacular palisades along Split Rock Creek and the Big Sioux River in Sioux Falls.

Sioux quartzite exists beneath 25 counties in eastern and central South Dakota, 11 in southwestern Minnesota, two in northwestern Nebraska, and at least one in northeastern Nebraska. Its maximum thickness is unknown, probably between 200 and 500 feet. The original extent of the Sioux quartzite is unknown; it looks very much like the Baraboo quartzite of Wisconsin.

Sioux quartzite at the Palisades, along Split Rock Creek, south of Garretson. —S.D. Department of Tourism

The formation is a massive pink quartzite that takes its name from outcrops along the Big Sioux River in South Dakota and Iowa. The rock consists predominantly of fine grains of quartz sand that owe their pink color to a thin coating of iron oxide. Quartz cement fills the spaces between sand grains.

Sioux quartzite rests horizontally on an erosion surface planed across much older metamorphic rocks. They are more than 1.65 billion years old. The oldest rocks on top of the Sioux quartzite are Cambrian sandstones that are less than 600 million years. The Sioux quartzite is the only formation in South Dakota that represents a portion of that great time gap of over 1 billion years. It is tentatively dated at about 1.2 billion years.

The quartzite consists mostly of thick beds of fine sand grains of very uniform size. A few thin streaks of coarse sand and gravel and a few layers of shale or clay within the quartzite range in thickness from an inch to as much as 30 feet. The small size of the sand grains and their uniformity both suggest that they may have been deposited in dunes, but the formation lacks the distinctive internal bedding characteristic of sand dunes. Both the clay and the sandstone contain ripple marks, which suggest they were deposited in shallow water. The clay shows mud cracks, which indicate that it was

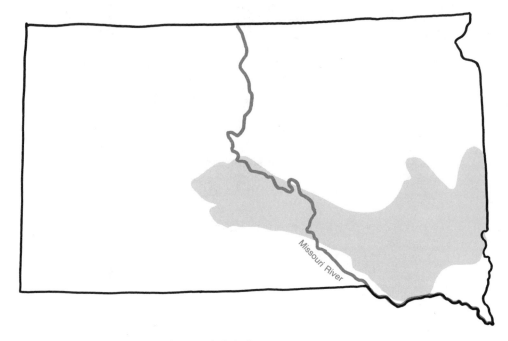

Missouri River

Area underlain by Sioux quartzite.

Sioux quartzite at Devil's Gulch, near Garretson, Minnehaha County. —S.D. Geological Survey

occasionally above water level long enough to dry out on its upper surface. Compaction turned the clay to pipestone, an indurated clay from which Indians carved pipes and ceremonial objects.

By the beginning of Paleozoic time, the main mass of Sioux quartzite was a steep mesa several hundred feet above the surrounding plains. Steep box canyons incised its edges. During Paleozoic time, the western end of the quartzite ridge sank. Later marine sediments gradually overlapped it. In Stanley County, just west of the Missouri River, more than 2,500 feet of Paleozoic and Mesozoic sedimentary rocks cover the quartzite. The eastern end stayed high, even above the widespread Cretaceous sea. Pleistocene glaciers finally overrode it, leaving deep striations where they dragged rocks across the quartzite surface.

Between the Minnesota State Line and Mitchell

Between the state line and Skunk Creek, just west of Hartford, the surface is old ground moraine deposited from the Illinoisan glacier. Several feet of windblown loess covers much of it. Wide meltwater channels filled with outwash sand and gravel entrenched the gently rolling glacial surface. Small streams now flow through the outwash channels, their short tributaries slowly dissecting and draining the moraine.

Palisades State Park is along Split Rock Creek, between Brandon and Garretson, where the stream cut through the glacial cover and entrenched itself into the Sioux quartzite. Rugged cliffs and a good growth of deciduous trees and shrubs provide an attractive setting for picnicking or camping. Take Exit 406 and follow a well-marked route to the northeast about 10 miles.

Minnesota line to Mitchell (82 miles).

33

Brandon is on a wide outwash plain deposited by Split Rock Creek and the Big Sioux River. Gravel pits abound.

Sioux Falls, the largest city in South Dakota, was established in 1858 on the banks of the Big Sioux River. Its founders recognized the potential for water power from the falls and the availability of an excellent building stone in the Sioux quartzite. To form the falls, the Big Sioux River cut through the glacial debris and into a buried ridge of Sioux quartzite. The normally placid river cascades over ledges of quartzite in a drop of at least 80 feet within less than a quarter of a mile. In the early days, a flour mill and numerous other smaller industries used the water power, which also generated electricity on a small scale. Most of the older, more substantial public and private buildings are built of dressed blocks of quartzite. More recently, the principal use of the rock is for riprap or as crushed rock for concrete aggregate and road metal.

Falls Park is just south of the stockyards. Foot trails permit a close view of the Sioux quartzite. At some prehistoric time, a glacial dam broke upstream, releasing a catastrophic flood. The torrent plucked huge blocks from the bedrock, creating a gorge with angular surfaces that contrast with the smoothly rounded surfaces of the quartzite eroded by normal stream abrasion. Take Exit 399 to visit the park.

I-90 continues on Illinoisan glacial debris almost to Skunk Creek. West of Skunk Creek, the glacial deposits are much younger and

Church built of hand-dressed blocks of Sioux quartzite quarried locally at Alexandria.

belong to the Altamont stage. The poorly established drainage, with numerous potholes and sloughs, is typical of young glacial deposits.

At the east fork of Vermillion Creek, between Humboldt and Montrose, the route passes from the poorly drained western edge of the Prairie Hills into the better drained, gently rolling topography of the James River lowland. Between the foot of the Prairie Hills and US 81, the highway is on still younger till left by the glacier of the Gary substage.

Between US 81 and Alexandria, the road crosses a poorly drained area full of potholes and sloughs. This is stagnation moraine deposited when ice of the Gary stage stopped moving and melted in place.

Two miles east of Alexandria, the highway crosses Pierre Creek. It flows in a channel eroded by floods of meltwater from the melting glaciers. Two miles south of Alexandria, the creek has cut through the glacial deposits, exposing the Sioux quartzite. This is the most westerly point at which the quartzite appears at the surface. Quarries here supplied much of the riprap used to prevent erosion of earthen structures along the Missouri River.

West of Alexandria, the surface is ground moraine, deposited from moving ice. The only notable feature between Alexandria and Mitchell is the valley of the James River. It meanders wildly on the floor of a trench cut into the bottom of the larger river valley. The catastrophic drainage of Glacial Lake Dakota eroded that inner trench, probably within a matter of a few days. It continues south to where the James River enters the Missouri River below Yankton.

Emery Meteorite

A polished slice of a stony-iron meteorite found near Emery is on display at the Museum of Geology in Rapid City. Stony-iron meteorites contain an alloy of nickel and iron woven through the mass of spherical stony chondrules. The polished surface nicely displays both kinds of material.

Stony-iron meteorites may be the primordial material of the solar system, condensed out of a glowing plasma of gas in space. In large planets like the earth, the original material has differentiated, with the metallic component sinking to the center to make the nickel-iron core. Meteorites of this type probably look like the material that aggregated to make the earth billions of years ago.

Between Mitchell and Chamberlain

The Mitchell Hills are low on the skyline 2 to 3 miles southwest of Mitchell. Before the glaciers came, they were a drainage divide between two preglacial streams. Bedrock there is the Niobrara chalk, thinly plastered with glacial debris.

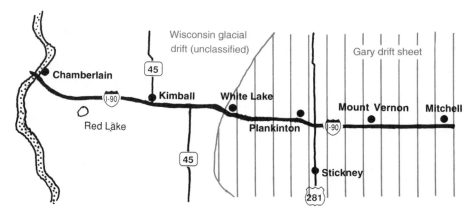

I-90, Mitchell to Chamberlain (70 miles).

A very gentle rise westward between White Lake and Kimball marks the change from the James River lowland to the more rolling topography of the Missouri Hills. The exact boundary is between Kimball and Pukwana, where the highway crosses a terminal moraine.

Red Lake, a short distance southeast of Pukwana, is a very shallow depression several miles across that partially fills with water in exceptionally wet years. It is probably a blowout excavated by the wind during a time when the climate was much drier than now, presumably during one of the interglacial times.

Corn Palace at Mitchell. —S.D. Department of Tourism

Bentonite beds in the Pierre formation, along Missouri River south of Chamberlain.
—E. P. Rothrock, S.D. Geological Survey

An overlook of the Missouri River at Chamberlain offers a good view of the Missouri River trench, which here is about 1.5 miles wide and 500 feet deep. The stream has cut down through several hundred feet of weak Pierre shale and nearly 100 feet into the underlying Niobrara chalk.

Imagine the scene before the river was backed up into the present reservoir. The Missouri River was a third of a mile across then, and groves of cottonwood trees lined its banks. The river flowed sluggishly, carried a heavy burden of silt, and constantly shifted the sandbars in its channel. Add a small sternwheeler with a flat bottom churning past the sandbars in a cloud of steam and wood smoke and you would have a reconstruction of the scene as it was before the railroads came. On its way upstream, the boat carried men and supplies for the Black Hills of Dakota, the gold fields of western Montana, and for all the little military posts along the river. On the downstream trip, it carried hides and furs, sacks of gold ore for the smelter at Omaha, and passengers returning to St. Louis for the winter.

The steamer *Far West*, which brought the survivors back to Fort Abraham Lincoln after the Battle of the Little Bighorn in 1874, was typical of the boats designed for navigation on the upper Missouri. She had a length of 190 feet, a beam of 33 feet, and three wood-burning boilers supplying steam to two engines. Empty, she drew only 20 inches of water. Fully loaded with 400 tons of cargo, she required at least 54 inches of clearance.

Chamberlain Wells

Before the valley was flooded to form Lake Francis Case, many artesian wells were drilled along the river bottoms, both above and below Chamberlain. Those wells were much shallower and had much higher surface pressures than those drilled on the adjacent uplands. An early well at Chamberlain encountered an artesian sandstone aquifer within the Dakota formation between depths of 633 and 689 feet. The well initially flowed a phenomenal 4,350 gallons per minute through an 8-inch casing. Its flow drove a small flour mill.

US 12
Big Stone City–Mobridge
210 miles

This route crosses the glaciated eastern half of South Dakota. Geographers divide the area into four distinct sections: From east to west, they are the Minnesota River lowland, the elevated Prairie Hills, the James River lowland, and the rolling Missouri Hills. Some consider the Missouri River trench a fifth unit.

Between the Minnesota Line and Milbank

Glacial Lake Agassiz formed when glacial ice blocked the Red River of the North in northern Manitoba. The lake became a shallow inland sea about 700 miles long and up to 300 miles wide. It covered most of southern Manitoba and parts of eastern North Dakota and northwestern Minnesota. Its overflow established an outlet to the southeast through the Minnesota River.

Torrents of glacial meltwater eroded that spillway into a trench about one mile wide and 150 feet deep; now, the little Minnesota River wanders on the broad floor of that great channel. In South Dakota, two long, slender lakes flood the northern end of the meltwater channel: Lake Traverse in the north and Big Stone Lake in the south. Lake Traverse is about 25 miles long, Big Stone Lake about 36 miles long.

US 12 crosses the meltwater channel at the south end of Big Stone Lake. The channel discharges into the Minnesota River through a

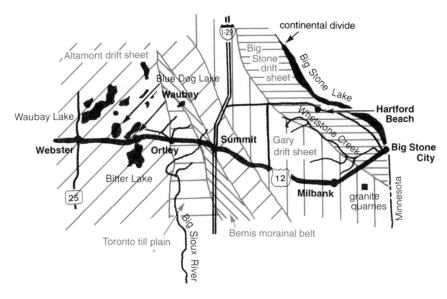

Minnesota line to Webster (57 miles).

spillway 964 feet above sea level, the lowest point in South Dakota. Lake Traverse reaches nearly to North Dakota. It drains northward, through the Red River of the North, ultimately to Hudson Bay. The low isthmus between the two lakes is the drainage divide between water flowing north to Hudson Bay and water flowing south to the Gulf of Mexico.

Mount Tom is 7 miles east and 2 miles south of Milbank. It is a knob of gravel at the north end of the Antelope moraine.

Sign for the continental divide.

Gold

Assays of a rock sample picked up near Big Stone City in 1889 showed traces of gold and silver. In those days, any trace of gold was enough to launch a gold rush. Miners even sank a shaft to a depth of 40 feet. Three years later, in the same vicinity, someone discovered traces of placer gold in glacial gravel. Of course, both the rock sample and the free gold were imported from Canada in glacial ice, presumably from somewhere in central Manitoba.

Milbank Granite

A coarsely crystalline dark red granite, the Milbank granite, lies beneath about 30 square miles of northeastern Grand County. Age dates come in at about 2.5 billion years. The Milbank granite consists of about 60 percent dark red feldspar, 25 percent clear quartz, and 15 percent black biotite mica. It crops out at the surface or lies under a thin mantle of glacial debris in a curving belt that reaches 3 to 5 miles east of Milbank. The first commercial stone quarry opened in 1908; the number of active quarries has since varied from one to eight. Variations in color from quarry to quarry have led to such trade names as Mahogany Granite, American Rose, and Royal Purple.

Under any name, the rough-hewn or sawn blocks of Milbank granite make an attractive building stone. The rock takes a high polish and is widely used as a facing on the lower portion of large buildings, as an interior trim, and in tombstones. You can see Milbank granite in the large stairway in front of the state capitol at Pierre, in Paul Bartlett's Statue of Patriotism at Duluth, Minnesota, and in the large columns in the National Catholic Shrine in Washington, D.C.

Quarry operations.

Quarry operations.

In late Cretaceous time, the inland sea advanced far enough east to submerge an erosion surface developed on the Milbank granite. Most of the sandy shale that collected on it was later eroded. Nevertheless, small pockets are preserved between the granite knobs in the vicinity of the Dakota Granite Company quarry east of Milbank. They contain the fossil remains of several species of sharks, sea turtles, and fishes.

Between Milbank and Summit

The route between Big Stone City and Summit crosses some of the youngest glacial deposits in the state. The last lobe of late Wisconsin glacial ice to invade this area from North Dakota and Minnesota divided at the northern end of the Prairie Hills. One part went southeast down the Minnesota River lowland; the other went down the James River lowland west of the Prairie Hills.

The east lobe tried four times to override the higher land to the west. The earliest, or Bemis advance, reached the crest of the Prairie Hills just west of Summit, where it deposited the long ridge of debris called the Bemis moraine. Then it melted back to the east. The second, or Altamont advance, reached the crest of the Prairie Hills, where it deposited the Altamont moraine. Then the Gary advance reached only to the foot of the Prairie Hills. The final, or Big Stone, advance stopped at Whetstone Creek, between Big Stone City and Milbank.

Esker, Grant County. —S.D. Geological Survey

The story is different on the James Valley side. The Bemis advance is absent; the Altamont advance overrode the Prairie Hills from the west, reaching as far eastward as Ortley. The Gary advance filled the James River lowland to the Missouri River at Yankton, but did not try to override the western side of the Prairie Hills.

The elevation of the glacial debris rises from about 1,100 feet in the Minnesota River lowland to more than 2,000 feet at Summit on the eastern rim of the Prairie Hills. Hummocky topography and many pothole lakes show that erosion during the past 10,000 years has done little to change the rough terrain or to establish a stream drainage pattern.

Prairie Hills

Between Andover and Summit, the highway crosses a rolling upland that the early French trappers called the Coteau des Prairies. Later settlers called it the Prairie Hills. By whatever name, the upland reaches nearly across the state from north to south and has an average width of about 50 miles. It stands as much as 900 feet above the Minnesota River trough, 400 feet above the James River lowland to the west. Broadly speaking, the Prairie Hills are shaped like a broad and shallow trough, higher at the edges than in the middle. The Big Sioux River system drains the low center south from near Summit to the Missouri River.

Much of the glacial debris that mantles the Prairie Hills is stagnation moraine. Such deposits develop near the end of a glacial stage when the ice stops moving and dumps its load of sediment as it melts. The result is a rough and poorly drained landscape full of small lakes, potholes, and sloughs. Although the Big Sioux River has established its main course, its tributaries still have not ramified into an integrated drainage net to drain the margins and the upper end of the Prairie Hills.

Between Summit and Aberdeen

The headwaters of the Big Sioux River are in older glacial debris, but ponds and swamps still survive in much of the area. The river channel is choked with gravel. Meltwater, carrying a huge load of debris, deposited the finer materials downstream, but left the coarser gravels behind. Ridges between Ortley and Waubay are terminal moraines deposited while the glacial lobe that came down the James River lowland remained stationary for a while.

Between Ortley and Webster, US 12 crosses a poorly drained area that contains several large lakes. They fill shallow depressions that were stream valleys before the glaciers came. Differential compaction of the much thicker accumulation of glacial debris that filled the old valleys left linear surface depressions over the old valleys. The highway crosses Rush Lake, which has an average depth of 6 feet and skirts the southern edge of Blue Dog Lake, which averages 8 feet deep. Short streams or sloughs connect them.

Waubay Lake and Bitter Lake, 3 miles to the south, have dissolved mineral contents of 7,000 and 25,000 parts per million, respectively. They are totally unfit for agricultural or domestic use. The salts accumulated because the lakes have no outlets.

The surface between Webster and Bristol is typical stagnation moraine, full of small potholes. Most formed where blocks of glacial ice buried in the glacial deposits finally melted, leaving roughly circular pits on the surface of the plain. These lakes do not connect at

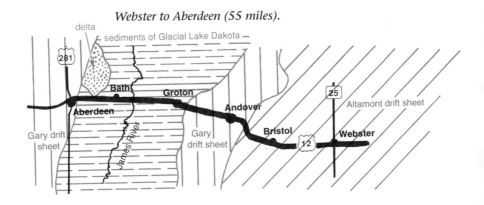

Webster to Aberdeen (55 miles).

the surface, although some may connect underground as water percolates through shallow outwash gravels.

Between Bristol and Andover the highway crosses between the Prairie Hills in the east and the James River lowland that extends westward for about 80 miles. In the James River lowland, the highway crosses a thin deposit of lake beds that cover an older ground moraine. In some places, an even thinner blanket of windblown loess covers the lake beds. The lake formed when meltwater backed up between the James River lobe of the glacial ice and the higher, older moraine to the east and south. It lasted only a short while and drained before Lake Dakota existed. Many smaller meltwater streams later dissected the lake bottom sediments.

The wide and swampy valley at the eastern edge of Groton is typical of the many channels that carried large volumes of meltwater as the last glacial front retreated to the north. By no stretch of the imagination could the present Mud Creek have carved such a wide valley.

The stretch of highway between Groton and Aberdeen crosses a flat lake plain of silty sediments deposited on the bottom of Glacial Lake Dakota. As the last ice sheet to fill the James River lowland melted, Glacial Lake Dakota backed up behind its terminal moraine, which functioned as a natural earthen dam. Silty sediment accumulated on its floor to a depth of about 50 feet. Much of that silt may have blown into the lake from nearby outwash plains. The very subdued highs and lows of the present surface probably express the surface of older ground moraine buried under that sediment. Numerous meltwater channels, which eroded after the lake drained, now incise the lake plain.

Many wells and test holes drilled in the Aberdeen area reveal the existence of a channel 200 to 300 feet deep carved in the old bedrock surface by torrents of meltwater as the glacial ice receded. It is now filled with sand and gravel. Bedrock consists of Cretaceous shales, chalks, and sandstones. They lie on a Precambrian granite basement at a depth of about 1,300 feet below the surface.

The Bath Meteorite

People watched in amazement as a meteorite that weighed 46 pounds fell near Bath in a blazing fireball on the afternoon of August 19, 1892. Searchers had no difficulty finding it on the silty surface of the sediments deposited in Glacial Lake Dakota, which are certifiably free of rocks. Most of the meteorite is stony, composed of chondrules about the size and shape of bird shot. It also includes some metallic nickel-iron. Several of the large museums in the United States have spectacular polished slices on display, as does the British Museum of Natural History.

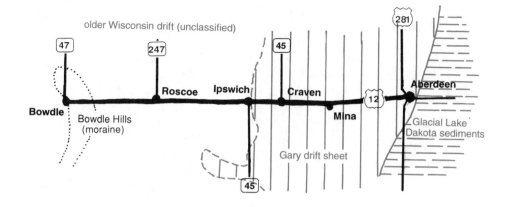

Aberdeen to Bowdle (52 miles).

Between Aberdeen and Mobridge

The highway is within the James River basin between Aberdeen and the area just west of Bowdle. The surface is mostly stagnation moraine deposited from ice of the late Wisconsin Gary advance. Some low ridges are probably terminal moraines, built up during periods when the ice front remained stationary. West of Ipswich, the surface drainage has not yet become well established.

A partial skeleton of a mammoth was found buried in the glacial debris on the O. C. Kienast farm 6 miles north of Mina. Mammoths died out at the end of the ice age, about 10,000 years ago. This short-haired mammoth stood about 13 feet high at the shoulders and had tusks 12 feet long. You can see it in the Museum of Geology in Rapid City.

The route between Bowdle and Selby crosses stagnation moraine less than 80 feet thick. West of Bowdle for 2 miles, the road is on a gravel outwash plain. The gravel covers an area of 14 square miles and averages 20 feet in thickness. Spring Lake probably occupies a depression formed over a buried segment of the pre-glacial channel of the Grand River.

Bowdle to the Missouri River (47 miles).

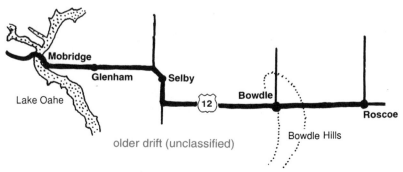

The Mina Mammoth. The original ivory tusks shattered when exposed to the air, but are faithfully reproduced in plaster of paris. —Phil Bjork, S.D. Museum of Geology

The highway between Selby and Glenham crosses stagnation moraine old enough to have begun to develop a surface drainage pattern. Blue Blanket Lake and the smaller lakes nearby occupy depressions that overlie small segments of the ancestral Grand River that flowed before the ice ages began. South of Blue Blanket Lake, US 12 crosses more than a mile of glacial outwash plain.

Ipswich is at the dividing line between the wide plain of Gary ground moraine to the east and the somewhat older glacial sediment that extends west to the Missouri River. Several buildings in Ipswich are made of glacial boulders, either as rounded field stones or dressed by skilled masons.

Many years ago, someone discovered southeast of Mobridge a glacial boulder that weighs about five tons and has two human hands etched on it, life size. A well-beaten trail led to the site. Archeologists think Indians may have placed their hands in the imprints while praying. The rock was moved to the public library grounds at Ipswich, care being taken to orient the stone precisely as it originally lay.

Just east of Bowdle is a prominent morainal ridge that extends more than 40 miles from north to south. It separates the James River lowland from the Missouri Hills to the west. Short streams drain the west side of the moraine to the Missouri River. About 10 miles southeast of Bowdle, a farmer was building a stock pond when he found nicely preserved white spruce wood and cones within this moraine, 12 feet below the surface. Radiocarbon dates showed that the wood was 12,220 years old, give or take 150 years. That is a surprisingly young date, right at the end of the last glacial advance. It helps correlate the age of the glacial deposits on the west edge of the James River lowland with those along the eastern edge of the state.

Most of the route between Glenham and Mobridge is in the breaks of the Missouri River Valley. Some patches of older glacial till and windblown loess partly cover the bedrock, dark Pierre shale. Scattered boulders, glacial erratics, rest directly on the shale, erosion having removed the softer glacial clays in which they were embedded.

The Missouri River trench formed when glacial ice dammed the eastward-flowing rivers that now drain only the western half of South Dakota. The impounded water worked its way south along the ice front, rapidly eroding the present channel of the Missouri River. Its floor is about 300 feet below the uplands at Mobridge. The steep sides of the trench permit countless small slumps and some very large landslides to develop within the weak Pierre shale. The result is a hummocky topography like that of the river breaks throughout the western half of the state.

Petroglyphs on a glacial boulder. Prayer Rock in the library yard at Ipswich.

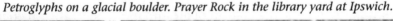

47

Mobridge is on a broad terrace within the Missouri River trench, near the upper end of Lake Oahe. Changes in water level within the reservoir make marked changes in the position of the shoreline, often exposing large areas of mud flats.

US 14
Minnesota Line–Pierre
220 miles

Glacial debris, left when the last ice sheet melted some 10,000 to 20,000 years ago, mantles the entire area. US 14 crosses the Prairie Hills from the state line to De Smet, crosses the James River lowland from De Smet to the vicinity of Miller, and crosses the rolling Missouri Hills from Miller to near Pierre, where the road drops abruptly into the Missouri River trench.

Between the Minnesota Line and Huron

The topography between the Minnesota line and Volga consists of older till plains dissected by wide and much younger outwash plains along the Big Sioux River and streams tributary to it. The Big Sioux River was a major drainageway for meltwaters from the late Wisconsin glaciers as they melted northward along the Prairie Hills.

Brookings is on a low plateau of ground moraine. Outwash channels, choked with deposits of sand and gravel, completely surround the town. The municipal water supply comes from several wells drilled 60 to 65 feet into the saturated outwash gravels. Between Brookings and Volga, the train of valley outwash next to the Big Sioux River is about 5 miles wide.

Minnesota line to Huron (95 miles).

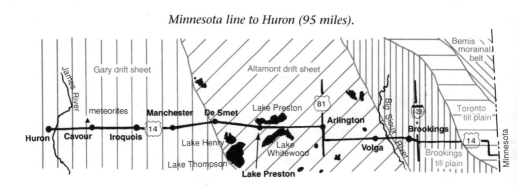

Elongate lake probably developed over an old pre-glacial valley by compaction of glacial till.

The land surface between Volga and De Smet is poorly drained, underlain mainly by glacial till dropped from stagnated glacial ice melting in place. Many shallow sloughs, ponds, and small lakes punctuate the ground moraine deposited by the Altamont ice advance during Wisconsin time.

East of the town of Lake Preston, the route passes between two such lakes, each about 6 miles long from east to west. Lake Whitewood is south of the highway; Lake Preston is north and is very shallow and usually dry. The famous explorer John C. Fremont named it during a reconnaissance trip through eastern South Dakota in 1839.

De smet is in a shallow outwash valley near the western edge of the Prairie Hills. Two miles west of town, the highway starts a slow descent of roughly 200 feet into the James River lowland.

The gently rolling surface between De Smet and Huron consists of ground moraine with a few low morainal ridges, a relic of the Gary glacial advance. The moraines trend from northeast to southwest, neatly delineating the southeastern margin of the last lobe of glacial ice to invade the James River lowland.

The James River meanders aimlessly in the floor of its trench, which is some 2,000 feet wide and 100 feet deep on the eastern edge of Huron. Obviously, that small river could not have eroded that broad valley. A catastrophic flood released when Glacial Lake Dakota washed out its morainal dam near Redfield eroded that trench, probably in a matter of days. Where US 14 crosses the James River, the stream impinges on the eastern side of the trench; most of the bottomland is now Memorial Park. Ravine Park Lake, on the north edge of town, is behind a dam at the mouth of a sharp ravine.

Wolsey meteorite, an iron meteorite 6 inches long.
—S.D. Museum of Geology

Meteorites

People found six stony-iron meteorites, aerolites, near Cavour between 1938 and 1944. They presumably fell in a single great shower, but no one knows when. In all likelihood, many more remain in the fields. They are worth seeking. Those that have been found weigh from 1 to 12 pounds. One fine specimen is on display in the Museum of Geology at the School of Mines in Rapid City. It weighs 8 pounds.

Bruce Eichstadt turned up a nickel-iron meteorite that weighs 160 pounds while cultivating a field near Wolsey in 1981. That is an absolutely spectacular find. A polished slice that weighs 22 ounces is on display at the Museum of Geology in Rapid City. It could not be part of the swarm of stony-iron meteorites around Cavour.

Between Huron and Pierre

The stretch of road between Huron and Wessington crosses a rather monotonous expanse of ground moraine. Many small sloughs and ponds show that the James River has not established an integrated drainage since the ice melted. A few low morainal ridges trend from northwest to southeast, precisely recording the southwestern side of the James River ice lobe of the Gary ice advance.

The surface between Wessington and Highmore is mostly ground moraine deposited as the Gary ice sheet stagnated and then melted. The general topography is monotonous, with a few very large shallow depressions. The town of Ree Heights lies in one such low area. The few subtle ridges do not appear to be moraines; they apparently formed through differential compaction of the glacial debris over

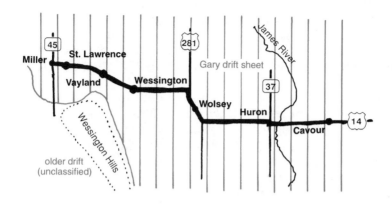

Huron to Miller (45 miles).

hills and valleys in the pre-glacial topography. Throughout this interval, the total thickness of glacial drift averages less than 200 feet. Drilling east of Miller has revealed two old valleys filled with more than 400 feet of glacial debris.

A gentle westward rise in the land surface near Miller marks the transition from the James River basin to the higher, more rolling Missouri Hills. The high area south of Ree Heights, which gives its name to the town, shows more clearly the break between the James River lowland and the Missouri Hills. The gentle change along the highway is scarcely noticeable.

In 1891, someone examined bottom sediment from a very small glacial lake 5 miles south of Ree Heights and found that the principal constituent of the deposit is diatomite, an accumulation of the siliceous skeletons of microscopic floating algae. It makes a good

Miller to Pierre (72 miles).

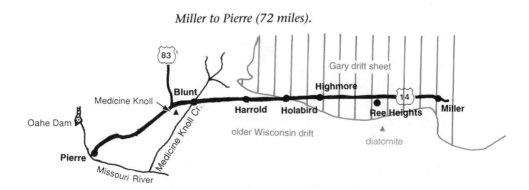

polishing powder. The deposits also contain many fossil fish, as well as the remains of snails, turtles, and frogs.

Highmore lies near the crest of a broad belt of low terminal moraines that marks the former western margin of a broad lobe of the James River glacier. Between Highmore and Pierre, the route crosses the Missouri Hills, where the surface is mostly ground moraine, accented by a few morainal ridges that trend from northwest to southeast. For several miles on either side of Blunt, the highway crosses glacial outwash along the floodplain of Medicine Knoll Creek, which flows southwest to the Missouri River. The highway cuts through a belt of moraines from 50 to 75 feet high. Gravel is produced from pits within the outwash.

Streams sort sediment according to size and deposit it in nice layers of clean gravel, sand, and silt. That is why glacial outwash makes such good raw material for a gravel pit. Melting glaciers dump their sediment as glacial till, a crudely layered deposit of unsorted debris that looks like something a bulldozer might have scraped together. A deposit of till is generally a very poor place to locate a gravel pit. After oil and gas, the sand and gravel business is the biggest mining industry in the country. So the distinction between outwash and till is not just an academic nicety.

The State Capitol used three kinds of building stones. —S.D. Department of Tourism

The highway between Blunt and Pierre crosses rolling glacial terrain. Geologists consider ridges less than 20 feet high to be irregular ground moraine. Those 20 to 40 feet high they map as terminal moraines. These apparently formed sometime during the middle of the Wisconsin ice age.

Medicine Knoll rises about 2 miles south of the junction of US 14 and US 83. It is a hill about 200 feet high made of Pierre shale and thinly covered with glacial debris. It lines up with the similar Snake Butte, 4 miles north of Pierre, and with Willow Creek Butte, 10 miles west of Pierre along US 14. The latter has a cap of gravel, derived in part from the Black Hills, which helps protect it from erosion. The three buttes may mark the path of an ancient stream that flowed east across an old erosion surface. If so, the last ice sheet probably removed the gravel cap from Medicine Knoll and Snake Butte.

As the highway approaches Pierre from the east, it gradually drops 300 to 350 feet into the Missouri River trench. Pierre is on several levels. The parkway along the Missouri River is on the present floodplain; older parts of town are on an alluvial terrace. The newer part of downtown is on a second, somewhat higher terrace of glacial outwash. Some of the newer housing developments are directly on Pierre shale higher in the breaks.

Several landslide problems have arisen in the areas where the Pierre shale is exposed. In some cases, people started old slides moving by excavating the slides' toes, or lower edges. In other cases, people caused ground movement by overloading higher parts of old landslides to make flat building sites.

Settlement of the Pierre area dates back to 1817, when Joseph La Framboise established a fur-trading post on the west bank of the Missouri River at the mouth of Bad River. With the start of the gold rush to the Black Hills from 1874 to 1876, Fort Pierre became the staging point for freight brought up the Missouri River by steamer. Ox trains then hauled it to the Black Hills. Pierre was merely the east end of the ferry across the river. With the arrival of the Chicago and Northwestern railroad in 1880, Pierre began to outgrow its neighbor. In 1883, Pierre tried to become the seat of the territorial government, but lost to Bismarck. In 1889, Pierre became the capital of the new state of South Dakota.

Pierre Gas Field

Early drillers encountered natural gas in shallow water wells drilled into the glacial deposits and Niobrara chalk. Then, in 1894, an artesian water well drilled into the Dakota sandstone at the Pierre Indian School produced gas. The gas is dry, which means that it is mostly methane.

No well has produced gas without water. In fact, the gas is dissolved in the water. The reduction in pressure as the water column approaches the surface lets the gas bubble out of solution. The source of the gas is probably plant material within either the Dakota formation or the enclosing organic shales.

As gas production increased, Pierre and Fort Pierre both installed gas street lights. A well at the Locke Hotel in Pierre furnished gas for both heating and lighting. Many ranches installed separators on their artesian wells and used the gas to heat homes and outbuildings. But the volume of gas is so limited that very little is now used. An artesian well on the State Capitol grounds still produces gas that may be ignited for the edification of visitors.

Artesian well on State Capitol grounds.

Pierre gas field, which yields natural gas in water from the Dakota sandstone.

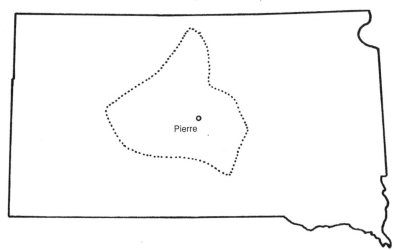

Oahe Dam

The Oahe Dam impounds the Missouri River about 6 miles above Pierre. It is a rolled-earth structure built on Pierre shale. Congress authorized its construction in 1944; construction started in 1948; and water storage began in August 1958. The dam is 9,300 feet long, 230 feet high, and contains about 90 million yards of earth. At full capacity, Oahe Reservoir stores about 23 million acre-feet of water. The reservoir extends north to within 5 miles of Bismarck, North Dakota, a distance of about 250 miles. Its functions include flood control, power generation, irrigation, and recreation. The name comes from the old Oahe Indian Mission established a short distance above the present dam in 1874.

Glacial boulders that eroded from the Canadian Shield and moved south in the ice were used as riprap at Oahe Dam.

Oahe Dam. —U.S. Army Corps of Engineers

Reactivation of old slumps and landslides caused considerable trouble at excavations at the dam site during construction. Riprap of glacial boulders picked up over a wide area north and east of Pierre protects the upstream face.

Iowa Line–Missouri River
116 miles

The last glaciers that invaded northeastern South Dakota melted back beyond the state boundaries a mere 10,000 or so years ago. Several tens of feet of debris, left when the ice sheet melted, cover the area from the Iowa line west to the Missouri River. It consists mostly of bouldery glacial till and outwash deposits of sand and gravel laid down by the torrential streams that carried away the meltwater.

Four glacial features make up most of the landscape along US 18. Over much of the area, the surface is a very gently undulating plain, underlain by glacial till dropped and smoothed off by the thinning, but still-moving, glacier. Occasional linear ridges are moraines that mark places where the ice front remained stationary for a while. Turkey Ridge, which is much higher than the moraines, has a rock core only thinly covered with glacial material. Old river valleys, choked with gravels dumped from the glacial meltwater streams, cut southward across this broad drift plain.

Between the Iowa Line and Olivet

The Big Sioux River is the boundary between Iowa and South Dakota. The highway bridge is about 1,250 feet above sea level. The higher, rolling land on the Iowa side is part of the Prairie Hills. On the South Dakota side is a lowland, roughly 200 feet below the Prairie Hills. The large hill north of the highway between the state line and Canton is an erosional remnant of the Prairie Hills. The Newton Hills, visible on the skyline 4 miles south of Canton, are another such remnant.

For many miles west of Canton, the highway crosses a monotonous plain of ground moraine underlain by roughly 100 feet of glacial till. Bedrock under the till is dark gray marine shales deposited during late Cretaceous time, when a shallow inland sea flooded most of South Dakota. You do not see those dark shales exposed until we approach the Missouri River breaks.

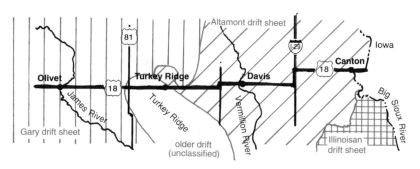

Iowa line to Olivet (56 miles).

Near the small settlement of Davis, the highway crosses the Vermillion River, which cuts into another outwash plain several miles wide. Farther west, Turkey Ridge Creek flows through another broad area of outwash gravels deposited from a glacial meltwater stream. If you are anywhere near a town, you can generally recognize outwash plains by the presence of active or abandoned gravel pits. Road metal and construction aggregate are major mineral resources, the basis for a large and profitable mining industry.

Turkey Ridge is the most conspicuous feature along this entire stretch of highway, standing as much as 400 feet above the prairie. The highway crosses it near its northern end. Before the great ice ages, Turkey Ridge was the divide between two large rivers. The glacier that came this way was too thin and weak to plane down the ridge, so it overrode it, leaving only a thin veneer of glacial debris. The rock core is Niobrara chalk, probably with some Pierre shale above it. Both were laid down in shallow seawater during late Cretaceous time. West of Turkey Ridge, the land is lower, forming the James River lowland.

The James River flows through a narrow trench eroded in the floor of a much broader valley, an arrangement that tells an amazing story. One of the great ice age glaciers left a moraine about 12 miles south of Redfield that impounded Glacial Lake Dakota, which was some 100 miles long and as much as 30 miles wide. The morainal dam finally broke, releasing a sudden flood that scoured the old James River Valley, cutting a channel as much as a mile wide and 100 feet deep—all within a few days. It reaches from the broken morainal dam to the Missouri River below Yankton, a distance of more than 200 miles. Where US 18 crosses the trench, you see a small stream meandering back and forth on its floor, a pathetic shadow of the catastrophic flood that once passed this way.

Olivet is on a remnant of an outwash plain between the James River and Lonetree Creek.

Between Olivet and the Missouri River

The route crosses monotonous ground moraine. You can distinguish in places two types of ground moraine, depending on the smoothness of the land surface. Where the surface is nearly smooth, glacial drift was deposited from moving ice. Where the surface is rougher, the ice was no longer moving, but melted in place. Near Tripp, some low ridges that trend southwest record places where the ice front remained stationary long enough to deposit moraines. They exactly delineate the margin of the vanished ice.

Lake Andes

This long and nearly linear lake occupies a valley eroded by glacial meltwater near the end of the last ice age. The lake, which now covers about 4,900 acres, has been set aside as a federal waterfowl refuge. Rainfall maintains the level of the lake with the help of two artesian wells drilled into the Dakota sandstone. Each is slightly less than 1,000 feet deep.

The hills south of the lake, next to the highway, are remnants of the Missouri Hills that farther north separate the James River basin from the Missouri River. Between the town of Lake Andes and Pickstown, the highway follows a pass about 2 miles wide where a meltwater stream broke through the high ridge to enter the Missouri River system. The sides of this cut expose Pierre shale, which locally is covered by only a thin mantle of glacial drift and isolated boulders.

Olivet to Fort Randall Dam (51 miles).

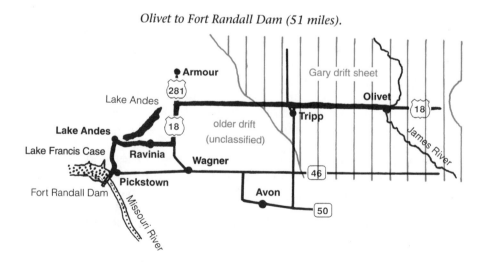

Pickstown was named for General Lewis Pick of the U.S. Army Corps of Engineers who helped devise the Pick-Sloan plan for development of the upper Missouri River basin. Under this plan, the Corps of Engineers and the Bureau of Reclamation jointly designed and built six dams across the main stem of the Missouri River between Fort Peck, Montana, and Yankton, South Dakota. They built the town as headquarters for the building and maintenance of the Fort Randall Dam.

Fort Randall Dam

This large, rolled-earth structure stands on Niobrara chalk and the overlying Pierre shale. Great chunks of white chalk excavated during construction of the outlet works and spillway face the dam.

Fort Randall Dam is nearly 2 miles long, is about 1,600 feet wide at its base, and stands 160 feet above the original floodplain. Construction started in 1946, and it began holding water in December 1952. At maximum pool elevation of 1,375 feet, it has a storage capacity of 5,700,000 acre-feet of water; at normal pool elevation of 1,365 feet, it has a storage capacity of 4,834,000 acre-feet. The power plant design accommodates eight generating units with a total capacity of 320,000 kilowatts. The dam backs the water up 150 miles, to about 60 miles above Chamberlain.

Fort Randall Dam. —U.S. Army Corps of Engineers

US 81
Yankton–Watertown
155 miles

Glacial till blankets the entire area of this route. The southern half, drained by the Big Sioux River, lies within the James River lowland. The northern part of the route crosses the somewhat higher and more rolling Prairie Hills.

Missouri River Trench

As it approaches Yankton from the south, the highway drops about 125 feet onto a broad, low terrace in the Missouri River trench. This

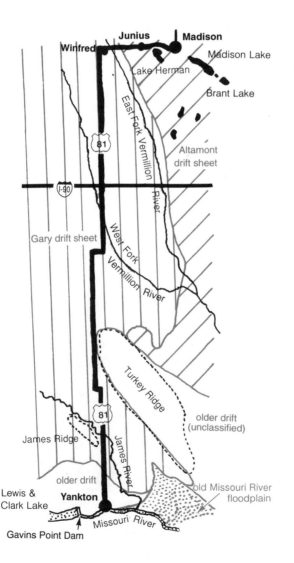

Yankton to Madison (94 miles).

remnant of an old floodplain stands about 40 feet above the present river level. The valley is about 2.5 miles wide at this point. The modern floodplain, covered with fine sand and silt, extends about a mile south from the Missouri River bridge. The river at this point has cut through the dark Pierre shale and deeply into the pale chalk and limestone of the Niobrara formation. A thin mantle of glacial drift and wind-transported loess hides most of the bedrock.

Gavins Point Dam

The U.S. Army Corps of Engineers built the Gavins Point Dam at a constriction in the valley. It is the farthest downstream of the series of dams that changed the Missouri River into a chain of lakes through eastern Montana and the Dakotas. Bedrock at the dam is Niobrara chalk; test drilling along the centerline of the dam showed that the depth of alluvial fill across the valley ranges from 35 to 70 feet below river level.

Gavins Point Dam is a rolled-earth structure about 9,600 feet long and 72 feet high. It was completed in 1955. A small powerhouse generates about 100,000 kilowatts of electricity. The dam backs the water up for about 37 miles. At pool level of 1,208 feet, the capacity of the reservoir is 477,000 acre-feet of water. The impoundment was named Lewis and Clark Lake for the explorers who described the rocks in the Yankton area in 1804.

Gavins Point Dam above Yankton. —U.S. Army Corps of Engineers

Yankton Area

At least five separate advances of glacial ice reached the Yankton area. Some crossed the present location of the Missouri River into northern Nebraska; others fell a few miles short of reaching the river. This ice was close to the southern edge of the great glaciers, so it was thin and moved sluggishly. It did not deeply erode the land. And when those glaciers melted, they laid down only a few tens of feet of glacial debris. Successive ice advances were too weak to bulldoze the deposits of previous glaciers, but overrode them, thus leaving a stack of deposits to record the glacial history.

Enough time passed between ice ages for soil zones to form on the most recent deposits of glacial debris. Then the next ice sheet buried the soil zone as it left its debris on top of the pile. The depths of those buried soil zones and the degree of weathering they show give a relative idea of the duration of the interglacial periods.

Ice advances in southeastern South Dakota are not yet well correlated with those in the upper Mississippi Valley. The total thickness of glacial deposits in the Yankton area averages about 50 feet, but is as much as 100 feet under the morainal ridges that stand above the general level of the prairie.

Glacial till overlies eroded surface on Niobrara chalk near Yankton. The 4-foot glacial boulder just left of the tree is limestone brought down from Canada. —H. E. Simpson, U.S. Geological Survey

Long after the glaciation ended, Yankton became the first capital of the Dakota Territory in 1861. In 1883, the capital moved to Bismarck. With establishment of North and South Dakota as separate states in 1889, Pierre became the capital of South Dakota.

Between Yankton and Madison

North of Yankton, the highway crosses a plain mantled with ground moraine deposited during the Gary ice advance. Two prominent ridges rise above it, with the James River flowing between them. West of the highway and about 10 miles north of Yankton is James Ridge. It is about 7 miles long and stands 270 feet above the prairie. Farther northeast is Turkey Ridge. It is about 40 miles long, 8 to 10 miles wide, and stands 400 feet above the surrounding flatlands. Both of these rocky ridges were stream divides before the ice ages began. Both consist largely of durable Niobrara chalk, with a cap of softer Pierre shale in some places. When the glaciers overrode these ridges, they left a thin veneer of till and gravel.

Thirteen miles north of Yankton, US 81 crosses the James River. Watch for its trench more than 100 feet deep and a mile wide. It was eroded catastrophically, probably in a matter of days, when Glacial Lake Dakota in the northeastern part of the state washed out its morainal dam. The little James River meanders sluggishly on the broad floor of the trench to its junction with the Missouri River at the east edge of Yankton.

Below Yankton, the Missouri River flows through an old valley. Notice that it is much wider than the narrow valley the river cut through central South Dakota after the ice displaced it from its original course.

Along several miles south of I-90, US 81 follows the breaks of the West Fork of the Vermillion River, which drains the southeastern corner of the James River basin. It joins the Missouri River at Vermillion. West of the West Fork is a poorly drained area underlain by ground moraine. Watch for the numerous pothole ponds and sloughs, great for ducks. The potholes probably formed as masses of ice buried in the moraine melted as the ice age ended.

Salem is the county seat of McCook County. Our route crosses a great buried ridge of Sioux quartzite. All Mesozoic and Paleozoic sedimentary strata are missing in this area, presumably because they were lost to erosion. The glacial debris lies directly on the pink Precambrian quartzite and has a depth of between 200 and 300 feet.

The absence of the usual sedimentary aquifers that lie beneath most of South Dakota greatly complicates the local problems of water supply. The Sioux quartzite contains virtually no open pore space, so it supplies no groundwater. The only remaining option is to drill wells into the glacial deposits. Glacial till is generally too impermeable to supply enough water to a well, so drillers hope to find outwash gravels below the water table. It is virtually impossible to find these gravels except by drilling.

North of Salem are two types of morainal deposits: The gently rolling and generally well-drained areas are ground moraine left by moving glaciers that smoothed the surface as they deposited the till. The poorly drained areas with sloughs and pothole ponds are stagnation moraines left where dead ice melted in place. Moving ice did not bulldoze that debris smooth. Both are deposits of the Gary glacial substage.

On the short east-to-west stretch of highway shared by SD 34 and US 81, the route crosses the East Fork of the Vermillion River. The east side of the river marks the boundary between the James River lowland to the west and the Prairie Hills to the east and about 100 feet higher. The Altamont ice sheet left glacial debris on the Prairie Hills. Both surfaces drain poorly and show abundant sloughs and pothole lakes. Lake Herman, on the southwestern edge of Madison, is a popular recreation area.

Between Madison and Watertown

The route crosses ground moraine and areas blanketed with outwash gravels left as the ice melted. Large lakes, such as Lake Poinsett, are in areas that geologists have interpreted as collapsed

Glacial lakes in Brush Lake Waterfowl Production Area, Brookings County. The lower pothole, now nearly filled with soil and vegetation, was originally deeper. —U.S. Fish and Wildlife Service

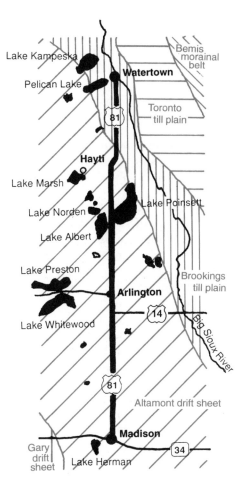

Madison to Watertown (62 miles).

outwash. As the ice sheet melted, outwash gravels covered large areas of stagnant glacial ice trapped in old valleys. When the buried ice finally melted, the gravels collapsed. If the ice was thick, a lake exists; if the ice was thin, only a slough may mark the spot.

Persistent low moraines mark places where the ice front remained nearly stationary for long periods. The hilly area between Hayti and Castlewood is part of the very large Altamont moraine that traces the old ice front for many tens of miles. Northeast of the Altamont moraine, the road passes onto the Brookings till plain, which it crosses to Watertown. The deposit is probably older than the Illinoisan ice age. Drainage is poor; many lakes and sloughs have formed in meltwater channels eroded as the ice melted.

US 83
Pierre–North Dakota Line
119 miles

After the highway climbs out of the Missouri River trench, it follows the Missouri Hills, a high and rolling country that separates the Missouri River trench from the James River lowland to the east. US 83 lies a bit west of the drainage divide, crossing short streams that began flowing west or southwest into the Missouri River after the last ice age.

Glacial ice sheets, moving slowly south or southwest, covered the entire area east of the Missouri River several times. The last glacier melted off eastern South Dakota about 10,000 years ago. That age comes from radiocarbon dates of samples of wood from trees that the advancing ice buried in glacial till. The road crosses deposits of bouldery glacial till and outwash sands and gravels left behind as the ice melted. Geologists have not yet established a firm correlation between the age of the till here and that laid down by successive ice advances along the eastern edge of the state.

Between Pierre and Blunt, the highway crosses two low ridges, moraines made of glacial till. They were deposited where the edge of the glacier remained stationary for a time. The edges of glaciers stabilize where the rate of ice advance balances against the rate the ice front melts. You see little evidence of glaciation between Blunt and Selby because the blanket of glacial material is thin.

In June 1954, the town of Onida drilled a well 2,111 feet to the Dakota sandstone. It flowed more than 400 gallons per minute and had a shut-in surface pressure of 255 pounds per square inch. The well behaved normally until February 1955, when water broke to the surface around the outside of the casing. It flowed wild for over a year, and disposal of the water became a major problem. Probing revealed a break in the casing 170 feet below the surface. Efforts to plug the well made the water again break to the surface 80 feet from the wellhead. A new well was drilled into the original well in August of 1959, and both were plugged with concrete.

Two early oil tests in Potter County reached the granite basement rock at about 1,800 feet below sea level. Neither found evidence of oil or gas in the overlying sedimentary rocks.

About 4 miles north and south of Selby, the road crosses stagnant ice topography. Watch for the irregular humps and hollows in glacial debris dumped from ice that had stopped moving and melted where it lay.

Except along about 5 miles at either end, US 83 crosses an outwash plain that roughly marks the position of a major pre-glacial valley. Geologists believe that much of this outwash was deposited on top of stagnant ice trapped in deep pre-glacial valleys. When the buried ice finally melted, the overlying outwash gravel subsided to form a rough, poorly drained surface with little or no surface drainage pattern. Undrained depressions, such as Sand Lake, probably mark places where the buried ice was unusually thick.

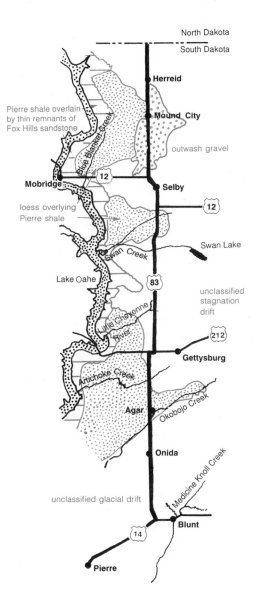

Pierre to North Dakota line (119 miles).

US 212
Minnesota Line–Missouri River
204 miles

Between the Minnesota line and the Missouri River trench, US 212 crosses four distinct physiographic provinces. From east to west are the Minnesota River lowlands, which extend into South Dakota only a few miles; the high Prairie Hills, which are about 60 miles wide; the James River lowlands, which are 90 miles wide; and the rugged Missouri Hills, which are about 30 miles wide.

Ice age glaciers covered the entire area. You see no bedrock, only bouldery glacial till, dumped directly from the ice, and clay, sand, and gravels deposited from meltwater streams. Most of the debris is ground moraine deposited from moving ice. A few morainal ridges formed at the margins of the glaciers. And valleys are filled with gravels left by streams that carried away the meltwater as the glaciers receded.

Between Redfield and Gettysburg, the route parallels a buried ridge of bedrock. Before the glaciers came, it was a section of the continental divide that separated streams flowing north to Hudson Bay from those flowing south through the Mississippi River to the Gulf of Mexico. The present divide lies much farther north.

Between the State Line and Clark

Just west of the state line, US 212 leaves the Minnesota River lowland and climbs several miles to the Prairie Hills. It rises steeply at first, then more gently. Near Tunerville, it crosses the crest of a glacial moraine where the land surface is 500 to 600 feet higher than in the Minnesota River lowland. The steeper part of this east-facing slope was originally unsorted glacial till, but the larger rocks have moved down the slope and collected in a rocky apron some 3 miles wide.

The Missouri Hills have the regional form of a gentle trough, highest near the eastern and western margins and lowest where the Big Sioux River bisects it on its way south. It resembles a metal platter upon which a roast turkey might be served, with a rim around the sides and a set of grooves down the middle that lead the juices to a sump at one end. The Big Sioux River has not had time since the end of the ice age to establish a completely integrated drainage system, so small potholes still remain back from the river on both sides.

Between Tunerville and Kranzburg, the highway crosses two strong belts of glacial moraines: the Altamont moraine to the east, and the slightly older Bemis moraine to the west. Most of the morainal ridges trend slightly west of north. Each records an exact position of the

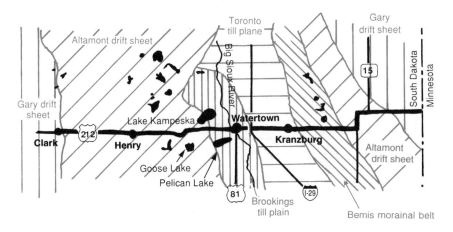

Minnesota line to Clark (64 miles)

old ice front. The highway reaches an elevation of 2,050 feet at Kranzburg.

Between Kranzburg and the area about 10 miles west of Watertown, the route crosses the Toronto till plain, a part of the Missouri Hills. Early glaciations left this long, north-to-south strip plastered with till, but later advances of the Wisconsin ice did not reach so far. Nevertheless, that later ice did shed torrents of meltwater that covered the old till plain with outwash gravels.

Watertown lies in an old glacial drainageway in a landscape of valleys choked with gravel, swamps, and lakes. The highway passes between Lake Kampeska on the north and Pelican Lake on the south. Gravel pits abound, both active and abandoned.

Between Watertown and Clark, US 212 crosses glacial till of the same age as the Altamont moraine on the east side of the Missouri Hills. It was deposited from a lobe of the same ice that moved down the James River lowland and encroached upon the Missouri Hills from the west. Here it is a stagnation moraine dropped from dead ice that was melting where it lay.

On the eastern edge of Clark is another glacial drainageway, this one poorly defined. Watch for the linear areas of swampy sloughs. Just west of Clark, the highway drops off the western edge of the Missouri Hills; the drop on the west is much gentler than that on the east.

Between Clark and Faulkton

The eastern half of this segment shows strong evidence of the manner of retreat of the tongue of glacial ice that invaded the James River lowland. The highway crosses many morainal ridges, all trend-

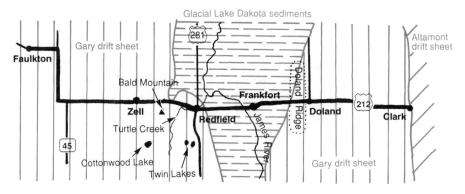

Clark to Faulkton (79 miles).

ing north to northwest. Each formed where the ice front stabilized for a few years before resuming its retreat to the north.

Doland is on Doland Ridge, a conspicuous moraine that continues for many miles on a north-to-south trend.

Along the 17-mile stretch east of Redfield, the highway crosses a very flat lake plain, once the floor of Glacial Lake Dakota. A glacial morainal dam, 12 miles to the south, blocked the drainage, impounding a lake more than 100 miles long and as much as 25 miles wide. As most morainal dams do, this one eventually washed out, and the lake drained in a catastrophic flood that eroded a channel south to the Missouri River at Yankton.

Several tens of feet of silty sediment accumulated on the floor of Glacial Lake Dakota. Watch for the gentle undulations in the present surface of that deposit; they probably reflect the old topography buried under the lake sediments. Water rushing past the broken morainal dam cut a few sharp drainageways into the old sediments.

Many of the early artesian wells drilled in the James River basin had phenomenal pressures. For example, a well drilled at Redfield in 1886 had an initial surface pressure of 177 pounds per square inch. For a short time, its flow drove generators that supplied electricity for the city.

The Gary advance of the Wisconsin ice age deposited ground moraine across the country between Redfield and Faulkton. The modern streams have developed an integrated drainage network, which explains the scarcity of sloughs and potholes.

Watch at 6 miles west of Redfield and 3 miles south of the highway for a large hill with a round top, Bald Mountain. It is an old bedrock hill of Niobrara chalk that acquired its streamlined shape when an advancing ice sheet overrode it.

The glacial deposits around Rockham lie in narrow and closely spaced ridges that are more easily seen from the air than from the car window. Some geologists call it a washboard moraine.

Between Faulkton and the Missouri River

The irregular surface west of Faulkton is a stagnation moraine. It is poorly drained and full of potholes and sloughs.

The Lebanon Hills are a moraine just east of Lebanon that trends from north to south. They are part of a long moraine that was originally correlated with the Altamont moraine farther east. This ridge separates the rough Missouri Hills to the west from the James River lowland to the east. West of the Lebanon Hills, the glacial deposits are older, and the ground moraine has a cover of windblown silt, or loess. Some still older glacial features form the low hills northeast of Gettysburg.

Between Gettysburg and the Missouri River, a thin mantle of old glacial deposits irregularly overlies the dark gray marine shales of the Pierre formation. Boulders scattered over the surface along the river breaks suggest that still-older glacial deposits have been eroded, leaving only these few durable relics.

The first wildcat oil well in South Dakota was drilled 2.5 miles northwest of Gettysburg in 1910. Drillers used cable tools to reach a depth of 2,260 feet. No oil.

Forest City Landslide

US 212 crosses the Missouri River on a 4,000-foot-long bridge built in 1958 when the Oahe Dam, 60 miles downstream, raised the water level at Forest City by 120 feet. Hummocky topography is apparent along the southern approach. The road is on a landslide block that is 8,000 feet long, parallel to the river, and 3,000 feet back from

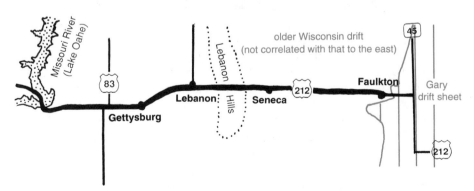

Faulkton to the Missouri River (59 miles).

71

the river. Rising water buoyed up the toe of the long-dormant slide, reactivating it. Irregularities appeared first in the 3,000-foot-long fill leading to the south end of the bridge. Then the whole mass threatened the south pier of the bridge. The moving mass of earth is largely developed within the weathered upper part of the Pierre shale, but up to 120 feet of glacial till and outwash gravel ride on top of the shale. Movement in the central part of the slide, which threatens the bridge, is a little more than 3 inches a year. Engineers moved great volumes of dirt from the new roadcut and loaded other parts of the slide, hoping to thereby stabilize the part threatening the bridge.

US 281
Nebraska Line–North Dakota Line
227 miles

In the short distance between the Nebraska line and the Missouri River, you can glimpse the flat sedimentary strata that lie beneath most of western South Dakota.

Just north of the Missouri River, the road enters the eastern half of the state, which once lay beneath a cover of glacial ice. There, the bedrock lies beneath a mantle of up to several hundred feet of bouldery glacial till and outwash sand and gravels. That debris covers a bedrock surface that once looked very much like what we now see in western South Dakota. The new, subdued topography reflects the history of deposition and partial erosion of this cover of soft and easily erodible blanket of glacial debris.

Between the Nebraska Line and Stickney

The highway enters South Dakota on an old prairie upland underlain by rocks made of pale sands and clays deposited during middle Tertiary time. The elevation is about 1,900 feet above sea level. The highway descends gradually as it follows ridges into the breaks of the Missouri River. The pale Tertiary clays that lie beneath the upland give way to dark gray marine shale of the late Cretaceous Pierre formation. In the valley bottom, at an elevation of about 1,250 feet, the Missouri River has cut entirely through the Pierre shale and about 20 feet into the underlying Niobrara chalk.

East of the Fort Randall Dam, the highway climbs out of the Missouri River trench. Watch for the glacial deposits mantling the outcrops of dark gray Pierre shale.

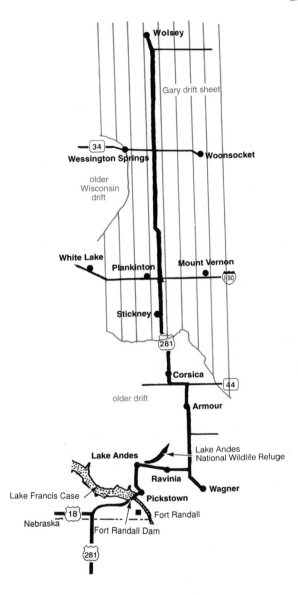

*Nebraska line to
Wolsey (115 miles).*

Between Fort Randall and Lake Andes, the route follows an old glacial meltwater channel eroded through the hilly ridges of the Missouri Hills. It drained glacial meltwater from the James River basin into the newly formed Missouri River trench. The highway follows the eastern edge of the old channel, which is about 2 miles wide and 200 feet deep.

East of the town of Lake Andes, the highway passes between hills of Pierre shale on the south and Lake Andes to the north. The lake is about 9 miles long, but quite narrow; it floods a continuation of the same meltwater channel.

The blanket of glacial debris that covers this upland is as much as 200 feet thick. Glaciers carried it down from the north, depositing it from a slowly moving ice sheet that reached as far south as the Missouri River. Where the glacier was still moving, it smoothed its debris into a nearly featureless ground moraine. But where it was stagnant, and melted in place, the stagnation moraine is much more irregular. Stagnation moraine extends north and south from Armour for several miles; the land is hummocky and full of potholes.

Between Stickney and North Dakota

Near Stickney, the highway crosses a poorly defined divide that separates drainages on the east that flow into the James River from those on the west that flow directly into the Missouri River. This divide marks the boundary between the elevated Missouri Hills to the west and the James River lowland that occupies much of east-central South Dakota.

Between Stickney and the North Dakota line, the highway lies within the James River drainage basin. The surface is glacial debris, ground moraine deposited by a single tongue of glacial ice that pushed its way down the James River lowland during the Gary substage of the last ice age. The area is still a hummocky, poorly drained plain, hardly changed since the ice melted some 10,000 or so years ago. The James River has been extending its tributaries, progressively setting up an integrated drainage pattern, but large undrained areas survive.

Between Aberdeen and the North Dakota line, the highway alternately crosses areas of gently rolling ground moraine and low morainal ridges that trend generally northeast. Torrents of meltwater, shedding from rapidly melting glacial ice, cut large valleys that now accommodate small creeks.

Glacial Lake Dakota

Between Redfield and Aberdeen, the highway crosses sediments that were deposited on the bottom of Glacial Lake Dakota. Any surface as almost perfectly flat as this must be an old lake bed. A few streams have slightly dissected it.

As the James River ice lobe melted, it stabilized for a while near the line between Spink and Beadle Counties, where it left a sizable moraine. When the ice front again began to melt back, meltwater collected behind the moraine, which made quite a good dam. This dam ultimately impounded a lake nearly 100 miles long and up to

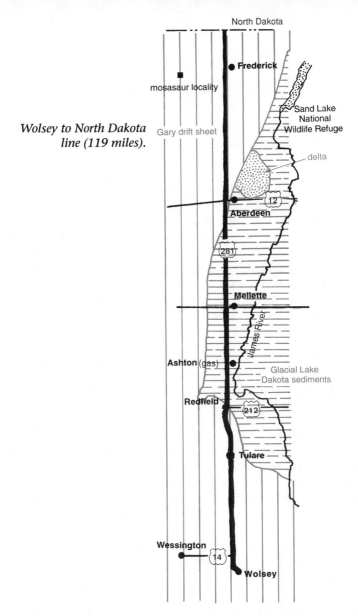

Wolsey to North Dakota line (119 miles).

30 miles wide—Glacial Lake Dakota. Silts and clays accumulated on the lake bottom. The thickness of this sheet of fine sediments ranges from a feather edge at the old shoreline to a maximum of 95 feet; the average is about 50 feet. Meltwater pouring into the lake built deltas of coarse sand and gravels at several points. Aberdeen is on such a delta. Pits southwest of town produce gravel for construction material and road metal.

Glacial Lake Dakota probably lasted for several thousand years. With luck, the dams that the Corps of Engineers build may last as long. Finally, the morainal dam washed out, and the lake emptied in a catastrophic flood that eroded a trench within the old James River Valley all the way south to the Missouri River at Yankton.

Wolsey or Eichstad Meteorite

A nickel-iron meteorite weighing 160 pounds was found near Wolsey in 1981. A polished slab, on display in the Museum of Geology at Rapid City, shows a rounded inclusion of the iron sulfide mineral troilite. No one knows when this meteorite fell, but is was presumably many years ago. A meteorite that size would make a spectacular fireball and all sorts of commotion as it passed through the atmosphere. Everyone in South Dakota would have known about it.

Although most of the meteorites that hit the earth are stony, most that people find are nickel-iron. They are easy to notice because they do not resemble ordinary rocks. Nickel-iron meteorites have essentially the composition of stainless steel, so they last a long time before they finally molder into rust.

Ashton Gas

A water well sunk at Ashton in 1881 encountered natural gas in glacial debris beneath the lake sediments. Decaying vegetation buried within the glacial clays apparently generated the gas, which for a few years heated a small hotel.

Frederick Fossil

In 1915 someone discovered the skeleton of a 29-foot mosasaur, a giant swimming reptile, in an outcrop of Pierre shale 10 miles west of Frederick. It is now on display in the Museum of Geology at the School of Mines in Rapid City. Mosasaurs were true reptiles, far more

Mosasaur. —S.D. Museum of Geology

closely related to the lizards than to the dinosaurs, who lived on land at the same time. Like the dinosaurs, they vanished in the catastrophic extinction that ended the Cretaceous period.

Pipestone, Minnesota–Pierre
223 miles

The route crosses four physiographic provinces, each a product of ice age glaciation. Pipestone is on the Prairie Hills, which extend 50 miles into South Dakota. West of that, the James River lowland reaches another 65 miles west. Farther west are the rugged Missouri Hills. The Missouri River trench is the western boundary.

Pipestone National Monument
The national monument at the north edge of Pipestone was established in 1937 to protect a quarry that the Indians worked more than 400 years ago. It is a good place to see the Sioux quartzite and outcrops of pipestone.

Sioux quartzite is a dense pink sandstone so solidly cemented that it breaks through the sand grains, rather than around them. It is as much as several hundred feet thick. The sand was deposited about 1.2 billion years ago in a shallow sea that covered at least eastern

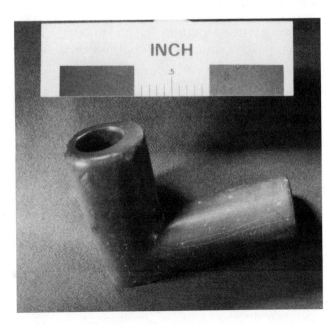

Indian Pipe carved from pipestone.
—Phil Bjork, S.D. Museum of Geology

South Dakota and nearby parts of Minnesota. The layers still lie nearly as flat as when they were deposited; evidently, this part of the earth's crust has been stable for more than a billion years.

Pipestone is a siliceous rock found in seams within the Sioux quartzite. It is typically pale to dark red in color, commonly with flecks of white. Pipestone is soft when first unearthed, but hardens rapidly upon exposure to air. Indians have carved it for centuries. It is also called catlinite, after George Catlin, an artist who visited the Indian quarries in 1832 as he painted his way through the upper Mississippi Valley.

Between the Minnesota Line and Egan

SD 34 crosses the state line on the broad Prairie Hills between the Minnesota River lowland on the east and the James River lowland 50 miles to the west. The last major advance of the ice was too weak to cover these hills. It split, sending one ice lobe east down the Minnesota River lowland, the other down the James River lowland nearly to the southern edge of South Dakota. Deposits of older glacial material exist on the Prairie Hills.

The road between the state line and I-90 crosses a long strip of prairie known as the Brookings till plain. It is underlain by a blanket of glacial till, sand, and gravel deposited during an earlier glaciation. The Big Sioux River carried tremendous volumes of water when the most recent glaciers were melting, greatly altering the till plain.

Egan is on gravels deposited in a wide valley cut by the ancestral Big Sioux River when it was carrying a maximum volume of meltwater. As the volume of water decreased, the streams were unable to carry their huge load of sand and gravel; they dropped the coarser material within their valleys. Some of these drainageways still contain small streams; others, such as the lakes near Madison, no longer contain active streams, but chains of shallow lakes and sloughs.

Pipestone, Minnesota, to Woonsocket (104 miles).

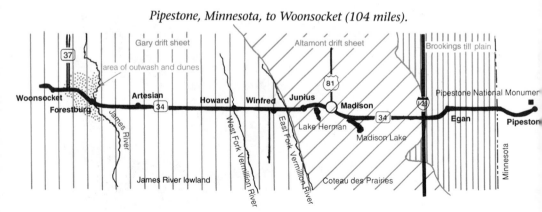

Flandreau Meteorite

A stony meteorite that weighed about 47 pounds was found near Flandreau in 1981. Although the real value of meteorites is scientific, they have also become collector's items. At the time of this writing, they were retailing at an average price of about one dollar per gram. At that rate, a meteorite that weighs 47 pounds would be worth a modest fortune. Meteorites do litter the landscape, and it is financially worthwhile, as well as interesting, to watch for them.

Stony meteorites are by far the most abundant, but are very hard to recognize. They are black rocks that feel a bit heavier in the hand than ordinary rocks, but not dramatically so. Most consist primarily of little spherical bodies called chondrules packed tightly together. Chondrules are no larger than a mustard seed and hard to see; it is impossible to identify them confidently except by examining a thin section under a microscope. Many stony meteorites also contain wisps of metallic nickel-iron woven through the dark mass of chondrules.

Between Egan and Woonsocket

Between Egan and Colman, the ice front remained stationary long enough to build morainal ridges. They are the same age as the Altamont moraine farther north, and they outline another former margin of the same glacier. The ice melted so recently, as such things go, that the streams have not yet extended tributaries to tap all the undrained depressions and sloughs.

A short distance west of Junius, the highway drops 200 to 300 feet to the James River lowland. The terrain is a rolling till plain still

Pothole lakes and former, now filled, pothole lakes mark stagnation moraine 4 miles east of Howard. Most of the area is mantled with a few feet of wind-deposited loess. —U.S. Fish and Wildlife Service

The James River flows placidly in its flood-formed inner channel south of Forestburg on SD 37.

much as it was when the Gary ice sheet melted more than 10,000 years ago. Low morainal ridges trend northeast across it. The highway crosses several streams that flow south now, as they did when they carried torrential floods of meltwater.

The town of Artesian was named for the many spectacular artesian wells drilled in the James River lowland in the 1890s. Flows were much larger than from wells drilled on the nearby hills because the water did not have to rise as high.

Jetted Wells

Thousands of early artesian wells, especially in eastern South Dakota, were jetted, an almost forgotten method of drilling. One end of a small-diameter pipe was flattened to make a cutting tool shaped like a chisel. The driller used a handle to rotate the pipe back and forth, scraping away the soft rock. Drilling-water pumped down the pipe returned through the annulus between the pipe and the wall of the hole, flushing the cuttings to the surface. The driller added segments of pipe as the hole deepened. A mast or a tripod-and-pulley arrangement was used to pull the drilling tools and set casing in the finished hole.

This method was remarkably fast and cheap in the soft sedimentary rocks of eastern South Dakota. Every farmer could have a well. Most wells on low ground flowed when they were new, then needed pumping as the water pressure dropped. Wells on higher ground

Jetting rig. —S.D. Geological Survey

generally needed a pump from the beginning. A South Dakota driller designed the mud circulation system used in 1901 to drill the first rotary oil well at Spindletop, Texas. The method is still in general use.

Between Woonsocket and Fort Thompson

The highway between Woonsocket and Wessington Springs crosses ground moraine deposited by the Gary ice sheet. A large area of glacial outwash extends for several miles north of Woonsocket. Prevailing northwest winds, winnowing this gravel, picked up fine sand and redeposited it in a minor dune field between Forestburg and Woonsocket. The sandy soil is ideal for raising watermelons and muskmelons. Forestburg melons are shipped for 300 or more miles.

Woonsocket stands on the divide between a large area of outwash gravels to the north and east, and hummocky, poorly drained morainal topography to the west. The town is famous among groundwater geologists as the site of a sensational artesian well of more than a century ago. It encountered the upper sandstones of the Dakota formation at a depth of 725 feet, then came in flowing 1,100 gallons per minute, and had a water pressure at the surface of 130 pounds per square inch. Another well, a few miles to the south near

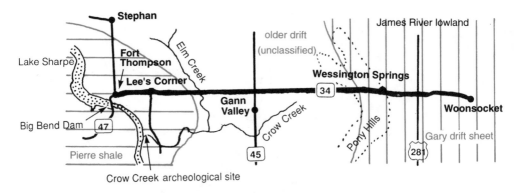

Woonsocket to Fort Thompson (59 miles).

Letcher, came in flowing 4,000 gallons per minute in 1890. Woonsocket has two newer artesian wells, about 750 feet deep, that furnish water to maintain a small artificial lake. It provides recreation and an emergency source of water for fighting fires.

Between Woonsocket and Wessington Springs, the road crosses ground moraine deposited by the tongue of ice that came down the James River Valley during the Gary ice advance. It reached the Missouri River at Yankton. As that ice melted, it left a series of low moraines. Here they trend northwest because they outline the southwest side of the ice tongue. Bedrock is Niobrara chalk.

Wessington Springs nestles at the foot of the escarpment at the eastern edge of the Missouri Hills. Leakage from a layer of water-saturated gravel, which crops out along the face of the slope, causes the springs.

Between Wessington Springs and Fort Thompson, the highway crosses glacial debris older than that in the James River lowland. The topography is more rolling, and the drainage has had more time to become established. Bedrock is Pierre shale except in the very bottom of the Missouri Valley, where the river cuts through it into the Niobrara chalk. Both formations were laid down under shallow seawater during late Cretaceous time.

The Pony Hills, near Wessington Springs, separate the James River lowland from the rougher country of the Missouri Hills to the west. Streams west of the divide drain into the Missouri River. The hills are glacial till dumped onto the eastern edge of the Missouri Hills. They drain poorly, in contrast to the plains of older glacial debris just to the west.

A monotonous blanket of ground moraine covers an area that extends some 50 miles west of the Pony Hills. West of Elm Creek, the terrain becomes rougher; the creek marks the approximate bound-

Early artesian well at Woonsocket. —U.S. Geological Survey

ary between the Missouri Hills and the Missouri River breaks. Dark outcrops of gray Pierre shale appear in the river breaks.

An Ancient Atrocity

More than 50 earthen lodges once stood on a point of land next to the Missouri River at the mouth of Crow Creek, 7 miles south of the highway. A deep moat protected the side of the village away from the river. In 1977, archeologists working in a ravine next to the village found the bones of more than 500 men, women, and children who had been slaughtered, scalped, and dismembered. The bodies were dumped in the ravine, and covered with clay. Then a fire was built over the burial. Radiocarbon dates on the charcoal

Big Bend Dam. —U.S. Army Corps of Engineers

show that it all happened about A.D. 1325. The area is now a National Historic Site. All of the bones were reburied at the site with appropriate ceremonies.

Big Bend Dam

Fort Thompson was established in 1863 as part of a program to relocate Indians who participated in the Sioux uprising in Minnesota in 1862. Construction of the Big Bend Dam required moving the settlement to a higher river terrace.

Big Bend Dam is a rolled-earth structure. Its base rests on Niobrara chalk; the ends abut Pierre shale. At maximum pool level, the reservoir impounds 1,725,000 acre-feet of water. The lake is named for

Fort Thompson to Pierre (60 miles).

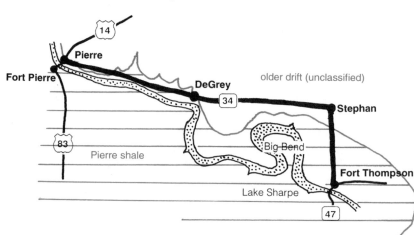

M. Q. Sharpe, a former governor who was a strong advocate of plans to develop the Missouri River and the manganese deposits in the Pierre shale.

Big Bend of the Missouri

The big loop that the Missouri River makes in central South Dakota has intrigued travelers since its discovery by the early trappers, who called it *le grand detour*. Several early explorers guessed at its dimensions. Lewis and Clark in 1804 sent a man to pace the distance across the neck, while the main party went around by boat. The man estimated the horizontal distance across the neck at 6,000 feet; those in the boat believed the distance around the loop was about 30 miles. Modern maps show that the distances are 8,500 feet and 22 miles.

Before the glaciers came, the streams that now drain western South Dakota continued east. Some went north to Hudson Bay; others drained south to the Gulf of Mexico. When the glaciers advanced across eastern South Dakota, they blocked those streams, backing

Satellite view of central South Dakota. Big Bend of the Missouri River is the huge meander in the lower right, Pierre is to the west at the narrowest part of the river, and the North Dakota line is just off the photo to the north. —EROS Data Center

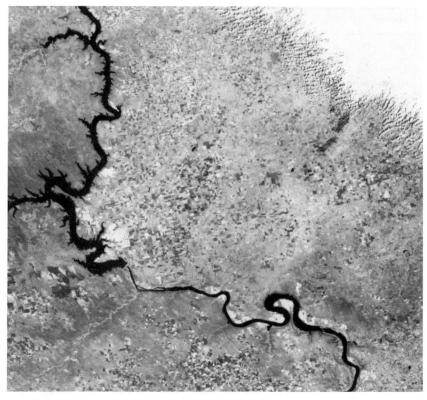

water into their valleys. Water rose against the ice front until it overtopped the low spots. Normally, the escape was to the south. At the Big Bend, the water found three valleys. It flowed northwest in one, southeast in the other two, ultimately carving the Big Bend loop. The only comparable feature is the Little Bend, at the mouth of the Cheyenne River, where the story is very similar.

For several decades, people talked of tunneling through the neck of the Big Bend to generate electric power. However, with a river gradient of less than 1 foot per mile, the project could not have generated enough power to justify the cost of the tunnel and power plant.

Between Fort Thompson and Pierre

The highway climbs out of the Missouri breaks and onto the upland surface just south of Stephan. It continues west on the upland for several miles, affording an occasional glimpse of the deep valley at the Big Bend. Near the western edge of the Crow Creek Reservation, the road drops back to the bottomlands and follows the floodplain of the Missouri River to Pierre.

Member		Description	Thickness in feet
Elk Butte		Noncalcareous olive gray shale and claystone with small calcareous concretions	160+
Mobridge		Calcareous shale; weathers to yellow and buff outcrops	100+
Virgin Creek		Olive gray shale and claystone; weathers to thin flakes; numerous thin bentonite beds	120–230
Verendrye		Light olive gray claystone, weathers to flakes and then to gumbo; some large calcareous concretions	140–160
DeGrey	shale-bentonite facies	Claystone and shale with persistent bentonite beds; manganese concretions abundant	135
	siliceous shale facies	Light gray siliceous shale; weathers to flakes; develops a stair-step outcrop pattern	125
Crow Creek		Marl, with thin siltstone bed at base	8–11
Gregory		Medium gray shale and marl	35–125
Sharon Springs		Dark gray-to-black shale with a series of bentonite beds from 1 to 18 inches thick	14–25

Members of the Pierre formation recognized along the Missouri River.

The defunct town of DeGrey gives its name to the DeGrey member of the Pierre shale. It makes a pale gray, normally bare outcrop littered with black concretions rich in manganese. The outcrops typically make a series of steps that owe their origin largely to many thin layers of bentonite.

DeGrey member of the Pierre shale, near DeGrey.

The highway enters Pierre on a low gravel terrace only a few feet above the floodplain. Pierre secures its water from a series of large wells sunk about 60 feet into the alluvial fill of the Missouri Valley on the modern floodplain.

SD 50

Junction City (I-29)–Lake Andes (US 18 and US 281)
100 miles

Much of the route east of Yankton is in the broad floodplain of the Missouri River—very flat ground. Farther west, where the Missouri Valley suddenly narrows, the highway climbs to the glaciated uplands, much dissected by many short streams that drain south into the Missouri River.

This segment of SD 50 follows the southern limit of Pleistocene glaciation in southeastern South Dakota. The ice front, controlled by the balance between the rates of ice advance and melting, constantly fluctuated. Sometimes it reached the position of the

Missouri River, other times it was many miles to the north. These fluctuations greatly increase the difficulty of sorting out the glacial history recorded in the glacial deposits.

Here, near its southern edge, the ice sheet was thin and moved slowly. It lay gentle on the land with little energy to plane off the high places. Rather, it rode over or around the older hills, stalling on some of the higher areas and melting in place. In lower areas where the ice was thicker, it continued to move, bulldozing its deposits smooth.

SD 52 offers an inviting side trip. It leaves SD 50 at Yankton and rejoins it a few miles farther west. This loop offers a chance to see spectacular cliffs of Niobrara chalk and the Gavins Point Dam.

Between Junction City and Yankton

The highway between Junction City and Vermillion crosses a gently undulating surface of glacial till in the form of ground moraine. Over most of this interval, deposits of windblown dust—loess—cover the till as a blanket several feet thick. Weathering of the loess has created a deep soil free of rocks, which is ideal farmland.

The Vermillion River flows past the western edge of Vermillion. Beyond the river, all the way to Yankton, the highway follows the broad floodplain of an ancestral version of the Missouri River. The soil consists of clay and silt dropped from muddy flood waters. Again, this is very deep soil, free of rocks.

The high hills north of Gayville mark the southern end of Turkey Ridge. It is about 40 miles long, 6 to 8 miles wide, and stands as much as 400 feet above the prairie. The bedrock is Niobrara chalk and Pierre shale, with a thin veneer of older glacial debris. Later Altamont and Gary ice tongues went around rather than over it.

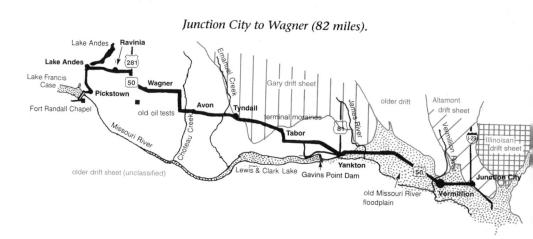

Junction City to Wagner (82 miles).

The highest point of Turkey Ridge, near its southern end, is Spirit Mound, so named because Indian tradition claimed that spirits guarded the area. A trapper, Charles LeRaye, visited Spirit Mound in 1802 and described it in his journal. Lewis and Clark made a side trip to Spirit Mound in 1804, and Clark described it in his journal.

The Yankton and Sioux City area was the dividing point between upper and lower Missouri River steamboat traffic. The river above Yankton required the use of smaller and more maneuverable craft than those used in the larger stream below. Boats plying the upper Missouri during the short open season frequently wintered and were overhauled at Yankton. Below Sioux City, the operating season was typically longer, and boatmen used larger craft. The heyday of steamboat traffic on the upper Missouri began with the discovery of gold in western Montana in 1862. It ended in 1881, when railroads reached Chamberlain, Pierre, Mobridge, and Bismarck. An early thaw in the spring of 1880 caused the river ice to break up suddenly, creating an ice jam at Yankton that damaged or destroyed many steamers moored there for the winter.

Both above and below Yankton, the Missouri River has cut deeply into the Niobrara chalk, forming nearly vertical cliffs in many places. A few miles downstream from Yankton, Lewis and Clark stopped at a chalk outcrop in August 1804. Clark recorded that "this bluff contained alum, copperas, cobalt, pyrites; an alum rock soft and sandstone. . . . " Most of what he saw was secondary minerals associated with the weathering of iron pyrite and its interaction with the calcareous chalk. The "cobalt," which he described as soft and transparent, was actually selenite gypsum, which abounds in such shale outcrops.

A bit upstream, in early September, Lewis and Clark found the bones of what they thought was a huge fish that they determined to be 45 feet long. Modern paleontologists believe they found the vertebrae of a mosasaur, a large marine reptile that terrorized the smaller denizens of the late Cretaceous inland sea. Mosasaur bones are common on both sides of the contact between the Niobrara chalk and the Sharon Springs member of the overlying Pierre shale.

For several hundred miles above Yankton, the Missouri River flows in a narrow gorge it eroded sometime during middle Pleistocene time. The glacier that then covered all of eastern South Dakota blocked the streams that normally flowed east. That forced the steams to flow southeastward around the ice margin, cutting a narrow gorge 300 to 700 feet deep through central South Dakota. After the glacier melted, the Missouri River continued to flow through the gorge, which was the lowest available route.

Quarry in Niobrara chalk at old cement plant. —S.D. Geological Survey

Yankton Cement Plant

Between 1891 and 1909, a cement plant operated just north of the present site of the Gavins Point Dam. Niobrara chalk and Pierre shale were blended, then fired to fusion in a large kiln. The fused clinker was ground to the consistency of flour, packed in sacks or barrels, and marketed over a wide area. During its 18 years of operation, the plant produced nearly 2 million barrels of cement valued at about $3 million. You can still see the ruins.

Between Yankton and Lake Andes

The road forsakes the river bottoms west of Yankton, but takes an easy grade to reach the upland surface 400 feet above lake level. The glacial cover is very thin near the head of the breaks; outcrops of Pierre shale abound in the short gullies that drop to the river.

Between the area just east of Tabor and that west of Avon, the highway crosses a belt of conspicuous glacial moraines that stand as much as 250 feet above the general prairie surface. These are terminal moraines deposited at the farthest reach of the ice. The ice front must have remained stationary for a very long time to dump such huge ridges of debris. Then the climate changed, and the ice melted back rapidly to the northeast.

Between Tyndall and Avon, the highway crosses Emanuel Creek, which divides the James River lowland from the Missouri Hills to the west. The creek was a major drainageway for great volumes of meltwater. Gravels in the old channel are a major source of groundwater. East of the creek is younger ground moraine; on the high ground to the west, the till is older, and was dropped from stagnant ice.

Between Avon and Wagner, the highway crosses the drainage basin of Choteau Creek. The surface is stagnation moraine dissected by many small tributary streams. Choteau Creek and its Dry Fork to the east were both important drainageways for glacial meltwater. Between Wagner and Lake Andes, the highway crosses ground moraine dropped from moving ice. Between Ravinia and Lake Andes, the highway crowds between the lake on the north and some hills of Pierre shale about 300 feet high that are lightly covered with glacial drift.

Wagner Oil Tests

Between September 1928 and July 1944, three wildcat exploratory wells were drilled within a very small tract 5 miles southwest of Wagner. The promoters sold stock during long intervals between the three projects.

All three holes were drilled with cable-tool equipment that pounded away for months on end, furnishing Sunday afternoon entertainment for people from a wide area. All three wells penetrated the Precambrian Sioux quartzite at about 1,450 feet, then drilled on into even older metamorphic rocks. The first hole continued to 2,320 feet, the second to 2,600 feet, and the third to 5,240 feet; it was for many years the deepest hole in the state.

Geologists of the time clearly understood that neither the Sioux quartzite nor the metamorphic rocks beneath it offered the slightest hope of oil production. The promoters must have understood that, too. It seems clear that they were in the drilling business, not the oil business.

Break in the Wall at the Jumpoff. Haystacks of Hell Creek formation are below the wall in the distance. —Phil Bjork, S.D. Museum of Geology

Rocks and Landscapes
West of the Missouri River

South Dakota west of the Missouri River contains four distinct geographic areas. The west-central part consists of rolling hills and stream valleys eroded in dark gray marine shale. They are partly covered by a thin veneer of windblown loess, stream bottom alluvium, and ancient stream gravels. The layers of rock lie almost flat. They are too weak to develop spectacular landforms.

Farther west, straddling the border between Wyoming and South Dakota, the rugged and darkly forested Black Hills stand several thousand feet above the surrounding prairies. They are more rugged than the Appalachian or Ozark Mountains. Although they are green when you are in them, they appear black from a distance.

Scattered highlands along the southwestern border of the state are eroded in layers of pale clays and sands laid down after the Black Hills rose. This blanket of clays once covered the entire region, but erosion removed most of it, leaving scattered remnants on the divides between streams. These remnants range in size from an acre or less to several square miles. We call them tables.

More widely scattered tablelands in the northwestern corner of the state are eroded in dark gray clays deposited during late Cretaceous time. Some have a cap of dark sandstone, others one of pale sandstone. These are called hills, rather than tables or mesas. Geographers refer to the area as the Cretaceous tablelands.

Some Cretaceous Formations

Everywhere in the world where sediments deposited in shallow seawater during late Cretaceous time exist, they seem to include the distinctive kind of limestone that people call chalk. Everyone has seen pictures of the white chalk cliffs of Dover in England. Their equivalent in South Dakota, is the Niobrara chalk. Like other late Cretaceous chalks, it consists largely of the minute fossils of microscopic animals that drifted in the warm surface waters of the shallow sea. The Niobrara chalk also contains the massive bones of mosasaurs and plesiosaurs, the great swimming reptiles that terrorized the late Cretaceous seas.

Above the Niobrara chalk is the dark gray to black Pierre shale. It is the most widespread and thickest sedimentary rock formation in South Dakota. It was deposited as mud rich in organic matter in a shallow inland sea that extended from the Gulf of Mexico to the Arctic Ocean. It once covered the entire state, except possibly the highest parts of the Sioux quartzite ridge around Sioux Falls. Erosion has now stripped it off the Black Hills and from smaller areas in eastern South Dakota. The Pierre shale was originally about 2,500 feet thick along the Wyoming state line and thinned gradually to several tens of feet along the eastern margin of South Dakota.

Where the inland sea was deepest, the appearance of the Pierre shale is very uniform. It is more variable near the eastern margin of the old seaway, east of the Missouri River. Variations are due primarily to varying quantities of lime and silica in the clay, which may record changes in the sediments brought in by the rivers entering the sea.

The huge area of the inland sea provided a niche for every form of marine life. Most of that shallow inland sea had a muddy bottom, which very few animals find congenial. That may explain why the most abundant fossils in the Pierre shale are ammonites, animals related to squid that swam in the open water and presumably cared little about the texture of the bottom. More than 80 species have been found in the Pierre shale.

Baculites compressus, *a common fossil in the Pierre shale, is related to the modern pearly nautilus. The intricate folded septa divide the shell into chambers. This section is 11 inches long.* —S.D. Museum of Geology

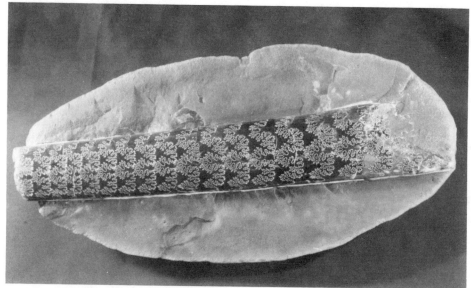

Here and there within the Pierre shale, little patches of resistant limestone erode into little conical hills called tepee buttes. The limestone formed from patch reefs that made hard areas on the bottom where animals such as oysters and clams could thrive. These buttes open a window into the vanished underwater world of late Cretaceous time. Most of the animals were invertebrates, most abundantly cephalopods, snails, and clams. Vertebrates include huge swimming reptiles, some as much as 40 or 50 feet long. Bird fossils are rare. Their fragile bones are not as likely to be preserved as are the more rugged bones of other forms.

Pierre shale was deposited over a period lasting about 15 million years. Very few animal species did not change during that period. Some species vanished, others evolved into new forms or were replaced by other species. Their comings and vanishings make it possible to divide the Pierre shale into fossil zones that probably represent periods of time. Fossil zones make it possible to correlate different kinds of rocks deposited at the same time, to trace the muddy shales of South Dakota into sandstones in western Wyoming or Utah.

The Cretaceous inland sea slowly retreated westward as the last of the Pierre shale was laid down, leaving behind a sequence of beach sands, mud banks, and coal seams—the Fox Hills formation. This assortment of coastal sediments was first studied and described in 1861 at Fox Ridge. The name was later extended to similar deposits throughout the northern Great Plains and northern Rocky Mountains. The Fox Hills formation ranges up to 200 feet in thickness. Where it is predominantly beach sands, they make an important aquifer that supplies water to many people.

After the Cretaceous inland sea retreated, rivers continued to import muds and sands, but spread them over a vast alluvial plain rather than dumping them on the coast. Streams wandered over a poorly drained landscape, shifting channels in flood time. During this final episode of Cretaceous time, up to 500 feet of dark clay, silt, and sand accumulated to form the Hell Creek formation. Locally, as in the Isabel-Firesteel area, beds of lignite coal several feet in thickness accumulated in coastal swamps. The Hell Creek beds are famous for their latest Cretaceous dinosaurs, including the triceratops with its three horns and the tyrannosaurus with its fearsome fangs.

The Flaming End of Cretaceous Time

Cretaceous time closed abruptly with the sudden extinction of approximately 65 percent of all the animal species then living. Many had existed for a long time, during which they appear to have competed quite successfully. The spectacular disappearance of the dinosaurs, the pterosaurs, the plesiosaurs, and the mosasaurs has caught public fancy, but they were not alone. The ammonite cephalopods, which had been a conspicuous part of the marine faunas all during Mesozoic time, disappeared, as did the microscopic animals whose minute shells accumulated to make the late Cretaceous chalks, and numerous other less famous groups of animals. What happened?

A great weight of evidence now suggests that an asteroid struck the earth 65 million years ago, causing environmental havoc on a global scale. Support for this hypothesis hinges, in part, upon a thin band of clay, the boundary clay, at the contact between Cretaceous and Tertiary sediments. It has been most closely studied in eastern Montana, but it also exists in western South Dakota.

The boundary clay is generally less than an inch thick, and it is exactly the same everywhere, on all continents and in the ocean basins. The only way to understand that uniformity is to conclude that it fell from the sky, like a fall of winter snow. Most geologists now interpret it as the dust cloud raised in the explosion of the impacting asteroid. In many localities, it contains enough soot to make it dark gray or black; evidently, great fires raged all over the earth as Cretaceous time ended.

Detailed studies in eastern Montana reveal a great variety of fossil pollen in the Hell Creek shales just below the boundary clay. It tells of fields of blooming flowers, many of them not very different from those we know today. No pollen exists within the boundary clay; evidently no flowers bloomed as the Cretaceous world crashed. The sediments immediately above the boundary clay contain a great abundance of fern spores, but no pollen. Then, the pollen returns, showing that most of the plants that lived before the catastrophe survived into Tertiary time.

Quite a few animals also survived the holocaust. They include all the general kinds of reptiles, such as snakes, lizards, and turtles. The birds also survived; many paleontologists include them with the theropod dinosaurs, the group that also produced tyrannosaurus. And most important from our point of view, the primitive little mammals that had scurried about in the underbrush of Cretaceous time also survived. They quickly evolved during early Tertiary time into a whole zoo of mammals, the inheritors of the land that the dinosaurs had so recently dominated.

Tertiary Time

Paleocene time, the first epoch of Tertiary time, began as Mesozoic time ended. The catastrophe that exterminated the dinosaurs had very little apparent effect on the physical environment of the northern High Plains. In extreme northwestern South Dakota, far from the rising Black Hills, continental deposition continued with no obvious break across the boundary between Cretaceous and Tertiary time. The Paleocene Ludlow formation lies directly on the Hell Creek formation and looks very much like it, except in being paler, less well cemented, and generally sandier. A total thickness of 350 feet was measured near Ludlow.

East of Ludlow and into North Dakota, the Ludlow formation contains layers of sandstone and shale that were deposited in shallow seawater—the inland sea again. Those layers belong to the Cannonball formation. It is a coastal deposit with fossils of small clams and snails that leave no doubt they lived in seawater. In some places, it is full of oysters. These few ledges exposed in scattered outcrops are all that remains to tell of the last time the inland sea flooded South Dakota. The Cannonball River takes its name from countless nearly spherical limestone concretions that weather out of the outcrops along its banks. The formation in turn takes its name from the exposures along the river.

The Tongue River formation, which lies above the Ludlow formation, is a thick sequence of sandstone and lignite coal best seen in the Cave Hills west of Ludlow. It is about 300 feet thick in western South Dakota and as much as several times that in eastern Montana and northern Wyoming, where it contains one of the world's largest reserves of coal. In northwestern South Dakota, the Ludlow formation contains one or two thick sandstone layers that make prominent brown cliffs. Clean sandstones, some deposited in estuaries, reach a thickness of 100 feet on several buttes and tablelands in Harding County. The Ludlow, Cannonball, and Tongue River formations constitute the Fort Union group in this area. The main source of sediments was the northern Rocky Mountains.

Eocene time was a period of tropically warm and wet climates, even in the high Canadian arctic. In South Dakota, it was a time of deep weathering and much erosion until near the end of the period. No sedimentary formations accumulated on land, evidently because rainfall was heavy enough to provide the streams with enough flow to carry all eroded sediments to the ocean. Weathering of the Pierre formation created red and yellow tropical soils.

Beginning in late Eocene time, about 40 million years ago, the streams again deposited sediments on land instead of carrying them to the sea. Evidently, decreasing rainfall deprived the streams of much of their flow. The new sedimentary deposits of the White River group were laid down all through Oligocene time and on through early Miocene time. They were deposited on late Cretaceous and Paleocene formations, generally burying the red and yellow soil developed on the Eocene erosion surface. The Oligocene sediments are pale clays and sands with a large content of volcanic ash.

Small remnants of the White River group survive on many buttes in northwestern South Dakota and in nearby parts of Montana and North Dakota. Most of those remnants are strongly cemented sands and conglomerates. The most complete section is in the Reva Gap area of the Slim Buttes, near SD 20, where between 500 and 600 feet of Oligocene sediments have been measured.

The Sharps and Monroe Creek formations lie above the White River beds. Together, they constitute the Arikaree group. The base of the Sharps formation contains a layer of reworked volcanic ash as much as 40 feet thick. Above that is more than 350 feet of pink siltstone, clay, and volcanic ash. Small concretions about the size and shape of potatoes are characteristic. The overlying Monroe Creek formation consists of up to 100 feet of wind-transported silt and volcanic ash, which probably erupted in the Western Cascades of Oregon and Washington.

Early Miocene time in western South Dakota was one of frequently changing conditions, probably easiest to credit to changing climates. Buried erosion surfaces separate six different sequences of Miocene sediments, six separate episodes of deposition. The lower two sequences are the most widely distributed and the thickest: The Harrison formation consists of about 150 feet of gray silty sandstone scattered through several counties in southwestern South Dakota. The overlying pink siltstone of the Rosebud formation is best displayed near Rosebud, Todd County, headquarters for the Rosebud Sioux. The four younger formations survive only in small outcrops. Each has its characteristic fossils.

During middle Miocene time, between about 17 and 15 million years ago, the entire region had a very warm and very wet climate. Red tropical soils formed from the Pacific coast all the way across the High Plains. Fossil leaves from that time in the Pacific Northwest are virtually identical to the plants that now grow along the Gulf Coast. A subtropical hardwood forest then covered the Pacific Northwest and may have reached east onto the High Plains.

Meanwhile, that wet climate and the rivers it maintained stripped a tremendous volume of Tertiary sedimentary formations off the High Plains and dumped the debris in the Mississippi Delta. In much of western South Dakota, the Tertiary formations were completely eroded, stripped right down to the Pierre shale. Scattered remnants of those formations remain in the high tablelands of western South Dakota to show what was lost. Most people see these tablelike remnants in the Big Badlands.

After the middle Miocene climatic interlude ended, the climate again became very dry. A new blanket of sand and gravel spread from the Rocky Mountains east across the High Plains, all the way from northern Alberta and Saskatchewan to Mexico and from the mountain front east to the central Dakotas.

That great deposit is called the Flaxville formation in Montana, the Ogalalla formation elsewhere. It contains abundant fossils of vertebrate animals, including early models of horses, camels, rhinoceroses, and elephants. Those old bones show that deposition began in late Miocene time, immediately after the wet climatic interlude, and continued until the end of Pliocene time, about 2 million years ago.

For some reason, western South Dakota contains less than its fair share of Ogalalla sand and gravel; the formation is much better developed in the neighboring states. It is not clear whether less was deposited in South Dakota or more has eroded off during the last 2 million years.

The pale sedimentary rocks of southwestern South Dakota offer a pleasant change from the drab marine shales that cover so much of the central and western parts of the state. These are the badlands, widely renowned for their spectacular erosional forms and their abundance of fossil remains of mammals and reptiles. They are a remnant of a blanket of similar sedimentary rocks that covered most of the western High Plains from Canada to the panhandle of Texas until they were eroded off when the climate became very wet during middle Miocene time.

Late Cretaceous, Paleocene, and early Eocene sediments, so widespread in northwestern South Dakota, are missing southeast of the Black Hills. Evidence is strong that they were deposited, then eroded rapidly during middle Eocene time. Studies of stream gravels of late Cretaceous and early Tertiary age west of the Black Hills in Wyoming show that streams carrying sediments from farther west flowed east across South Dakota, dropping their loads of sand and mud along the way. When the Black Hills started to rise, those streams were diverted around their northern end, and the sediments were rapidly eroded. Farther from the Black Hills, the Cretaceous shales weathered into deep red and yellow soils.

As the climate dried during late Eocene time, streams were no longer able to transport their burden of clays and sand. Deposition on land resumed and continued through Oligocene and early Miocene time, laying down hundreds of feet of pale strata. The color is due primarily to vast volumes of volcanic ash, wafted eastward from the Pacific Northwest, and mixed with the stream sediments. During a few million years of middle Miocene time, the climate became very wet, and most of the pale sediments were eroded off the landscape. Except for a few patches of stream gravels, no deposits of Pliocene age exist in southwestern South Dakota.

Road Logs West of the Missouri River

I-90
Chamberlain–Rapid City
211 miles

Except for a few miles at either end and an interval between Wall and Kadoka, this stretch of more than 200 miles is on the nearly horizontal Pierre shale. Because of the softness and homogeneity of the shale, erosion created few specific points of scenic or geologic interest. So, we will stress engineering features associated with the shale and the water resources upon which the region relies.

A Curse of Shale

The bumpy and undulatory pavement across western South Dakota emphasizes the difficulties engineering geologists and civil engineers have in building structures of earth or concrete on the Pierre formation. The shale has a high content of the swelling clay mineral montmorillonite, often called bentonite. So the Pierre shale swells when it is wet and shrinks when it dries.

Such shales are typically nearly impervious. However, water can and does enter the upper few feet of the shale through dehydration cracks, animal burrows, and the numerous small faults that formed when the original soft clay compacted into shale. During highway construction, engineers make every effort to provide proper drainage and to excavate and refill unstable areas, but they cannot anticipate all such problems.

As you drive along I-90, notice the bumps and the patched or replaced areas of pavement. Most of these failures relate to one of at least four problems. First, where the road passes from fill onto a rigid concrete structure, such as an overpass, the fill not only changes volume with the seasons, but in many instances the structure itself bulges up as the clay under the footings swells. Second, where the road enters and leaves roadcuts, buckling of the pavement occurs at the transition from fresh shale in the bottom of the cut to weathered shale on the old slope and between the weathered shale and the compacted fill of the roadway. Third, in open areas, the pavement may buckle at zones of minor faulting where water saturates the subgrade each season. Fault zones may be indicated by changes in vegetation in borrow pits or adjacent fields. Fourth, the Pierre shale is very unstable on steep slopes, and surface creep and landslides are common. Most earth movements occur in the spring, when

Highway landslides on Pierre shale. —S.D. Department of Transportation

the surface shales are saturated with water. Landslides are common along highways built on steep hillsides. The lower edge of the grade frequently fails, and entire sections of highway may move downslope overnight.

Water entering the shale also becomes highly mineralized, attacking and greatly shortening the life of metal culverts. Even the life of concrete structures may be seriously shortened if they are in contact with the more corrosive parts of the Pierre shale.

Between Chamberlain and Reliance

The Missouri River was an early barrier to east-west travel, except in the winter when teams and cattle could be driven across on the ice. A cumbersome pontoon bridge built at Chamberlain in 1893 interfered with steamboat travel. The first railroad bridge was built in 1905, the first highway bridge in the 1920s.

American Island, midstream opposite Chamberlain, was a popular park and campground until the Fort Randall Dam flooded it. When Lewis and Clark passed this point on their way up the river in 1804, Captain Clark, with his usual disregard for spelling, noted "a large perportion of seeder trees" upon the island.

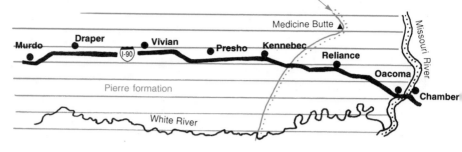

western limit of glacial erratics

Medicine Butte

Murdo Draper Vivian Presho Kennebec Reliance

I-90

Pierre formation Oacoma Chamber

Missouri River

White River

Chamberlain to Murdo (74 miles).

The I-90 bridge spans the original river as it was before the Fort Randall Dam raised the water level. Big blocks of pink Sioux quartzite protect its causeway across the old floodplain against wave erosion. From the west end of the causeway to Oacoma, the road rests on a low terrace left as the river eroded its channel to a lower level. Rock under the alluvium is Niobrara chalk; you may see it cropping out along the lower few feet of the river breaks.

Between Oacoma and Reliance, the highway follows the gentle gradient of American Crow Creek for many miles. Bedrock is Pierre shale.

At Iona, 26 miles south along SD 47 from Exit 251, the Pierre shale contained the skeleton of the giant swimming reptile *Alzadasaurus pembertoni*. Personnel of the Museum of Geology at Rapid City collected the bones and prepared the specimen in 1945. It is 36 feet long, a real monster of the late Cretaceous inland.

Rippled Sioux quartzite in riprap at west end of I-90 bridge at Chamberlain.

Niobrara chalk cliffs at west end of I-90 bridge at Chamberlain are nearly submerged by water behind dam.

From the upland surface at Reliance, you can see Medicine Butte with its transmission tower 6 miles to the north. The butte stands 450 feet above the prairie. It is one of many widely scattered remnants of an earlier land surface now almost totally lost to erosion.

Watch as far west as Kennebec for occasional weathered granite boulders resting on the shale surface. You do not see them farther west. They show that an earlier glacier extended to Kennebec before the Missouri River eroded its deep trench sometime during middle Pleistocene time.

Plenty of Manganese, but None to Mine

Geologists have followed a continuous dark band, about 50 to 80 feet thick, within the Pierre shale along the Missouri River from the middle of the state to the Nebraska line. They call it the manganese zone. It owes its color to countless purplish black concretions about the size of potatoes that constitute as much as 7.5 percent of the shale. They contain from 4 to 25 percent manganese.

Manganese nodules form today on enormous areas of the deep ocean floor in regions so remote from land that they receive very little sediment. These in the Pierre shale are generally similar, but they could not have formed in very deep water, and this region was not all that far from the coast of the Cretaceous inland sea. No one seems to know exactly how the modern manganese nodules form, and neither does anyone clearly understand these in the Pierre shale.

Manganese nodules composed of calcium-iron-manganese carbonate. They are light gray below the weathering zone, but rapidly oxidize to purplish black.

Manganese is an essential raw material used to make many kinds of alloy steels, especially high-speed tool steels. No industrial economy can survive without manganese. We import most of our supply from Russia and Brazil. It would be nice to have a large manganese mine in North America.

Between 1941 and 1947, the U.S. Bureau of Mines mapped, sampled, and tested the South Dakota manganese deposit at a site north of the highway, at mile post 257.2. They experimented with methods of mining and separating the nodules. Those tests revealed that environmental problems and the high cost of separating the manganese would make it impossible to mine the deposit except in the direst national emergency. Even so, it is nice to know that this enormous reserve exists.

Between Reliance and Wall

West of Reliance, the highway crosses a gentle divide, then drops into the valley of Medicine Creek, which it follows faithfully to Draper. Between Draper and Kadoka, it follows the high divide between the Bad and Cheyenne Rivers. West from Belvidere for about 6 miles, the highway is on a thin remnant of pale badlands clays of the White River group. They lie on the dark Pierre shale.

Kennebec takes its municipal water supply from a small artificial lake. Presho, Vivian, and Draper have artesian wells 1,500 to 2,100 feet deep. Water for Murdo, and for the two rest stops west of town, comes from 2,300-foot wells in the Cretaceous Dakota sandstone. Artesian pressure brings the water to within about 200 feet of the surface.

The relationship between the artesian sandstones in the eastern part of the state and those that crop out around the Black Hills remained unclear until the city of Kadoka drilled a pioneering deep

104

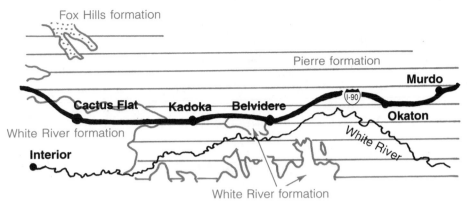

Murdo to Cactus Flat (54 miles).

artesian well in 1950. It showed two distinct thick artesian sandstones separated by a thick wedge of black marine shale. The upper or Dakota sandstone is of a big sand delta built by rivers entering the eastern side of the Cretaceous inland sea from the highlands of Minnesota and Wisconsin. Deposits in the delta thin westward; near the Black Hills, they show only channels of sandstone 20 feet or so thick called the Newcastle formation. The lower sandstone becomes progressively younger as it approaches the eastern shore of the Cretaceous sea. The wedge of Skull Creek shale, 250 feet thick around the Black Hills, thins eastward into its equivalent sandstone east of Reliance. In both aquifers, water quality varies greatly in the west-central part of the state. At Kadoka, the Dakota sandstone water is better; at Wall the deeper water has a lower mineral content and a higher artesian head.

Wall is at the extreme eastern edge of a large tableland. The town name comes from the Badlands Wall, a cliff some 200 to 300 feet

Cactus Flat to Rapid City (73 miles).

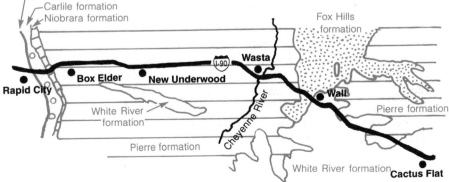

high that extends south for many miles. In the early days, it was a formidable barrier to travel. The town grew where the railroad scaled the wall when it came through in 1907.

In the early days, Wall depended upon a small artificial lake southwest of town for its water supply. In 1960, the city drilled the first of several 3,200-foot wells into the Lakota sandstone and abandoned the surface source.

Between Wall and Rapid City

Three miles west of Wall, I-90 starts a gradual descent down the Bull Creek valley into the 500-foot-deep valley of the Cheyenne River. The hummocky valley sides tell of countless small landslides in the Pierre shale.

The transitional contact between the Pierre shale and the Fox Hills formation is near the top of the bluffs to the northeast, 75 feet below the base of the light band at the top of the outcrop. This is the most southerly remnant of the Fox Hills formation, which is so widely distributed in the northwestern quarter of the state.

Just west of Wasta, several feet of sand and gravel cap the upland surface directly on the nearly impervious Pierre shale. Rain and snow meltwater sinks through the gravels, but stops at the shale. The water then moves slowly eastward through the gravel to the bluffs along the Cheyenne River, where it escapes as numerous small springs. Wasta gets a modest supply of very good water from shallow infiltration wells in the high terrace gravel. Pumps raise the water to a nearby storage reservoir, whence it flows downhill into town.

Alluvium in the Cheyenne River bottom furnishes most of the sand used for construction in western South Dakota. Pits are at accessible points as far west as the Wyoming state line. A small pit and washing plant at Wasta have been in operation for several decades.

West of Wasta, the highway climbs out of the Cheyenne River trench by following a narrow divide between two sharp tributary creek valleys. Engineers recognized the potential instability of this interval and originally used asphalt paving rather than the more rigid concrete.

Before 1951, the people of New Underwood relied for water on individual wells in the floodplain gravels along Box Elder Creek. Then they built an infiltration gallery in the floodplain and installed a treatment and distribution system. The infiltration gallery did not produce quite enough water, so the town drilled a well 2,700 feet deep to the Lakota sandstone in 1961 and a second, slightly deeper, well in 1988.

Between New Underwood and Ellsworth Air Force Base, the view to the south shows a rugged escarpment on the skyline 2 to 4 miles away. This marks the abrupt northern edge of a gravel-covered plateau between the Box Elder and Rapid Creek drainages. Other such surfaces separate other pairs of drainages east of the Black Hills. All slope gently down to the east and to the south. That explains why the northwest end of the long runway at Ellsworth Air Force Base is several tens of feet higher than the southeast end. Ellsworth Air Force Base is on one of those high surfaces on the divide between Elk and Box Elder Creeks. Pierre shale underlies as much as 30 feet of terrace gravel.

Those high and gravelly surfaces between the streams must be remnants of a formerly continuous surface on which the modern streams began to flow and into which they entrenched their valleys. The veneer of gravel leaves no doubt that running water once flowed across them. The only modern erosion surfaces that fit the description are desert alluvial plains like those that slope away from the flanks of the mountain ranges in Nevada. Fossil bones found in the gravels date from late Miocene and Pliocene time, as recently as 2 million or so years ago. Evidently, South Dakota was a desert then. The modern streams began to carve their valleys into that old desert surface about when the time of the great ice ages began, no doubt because the climate became wetter then.

Box Elder grew like Topsy when Ellsworth Air Force Base was built in 1941. It had no integrated water system until 1965, when the town drilled a 2,200-foot well to the Lakota sandstone; another was drilled a few years later. Those wells did not produce enough water, and what they did produce had a poor flavor. In 1983, the town drilled yet another well to the Madison limestone at a depth of 4,483 feet.

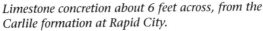

Limestone concretion about 6 feet across, from the Carlile formation at Rapid City.

Between Exits 57 and 60, the bedrock strata begin to tilt down to the east as they rise toward the Black Hills. Within about 3 miles, the highway cuts across about 1,500 feet of sedimentary rocks. It successively cuts across the upturned edges of the Niobrara chalk, the Carlile shale, the Greenhorn limestone, and the shales of the Graneros group before reaching the Cretaceous sandstones in the gap that I-90 follows into the Black Hills.

At Exit 59, the road crosses a ridge held up by a zone of huge limestone concretions in the Carlile shale. Individual concretions reach diameters of 5 to 6 feet. Their interiors apparently shrank after the outside hardened, opening fractures now filled with crystals of golden calcite. This zone of concretions continues for many miles into eastern Wyoming and Montana. In some areas, the zone contains abundant and nicely preserved remains of a variety of shellfish.

US 12
Mobridge–White Butte
110 miles

The highway climbs steadily from 1,600 feet above sea level at Mobridge to 2,600 feet at White Butte. It also moves up the sedimentary rock section, from the late Cretaceous Pierre shale at Mobridge to the Oligocene White River clays at White Butte. The sequence of formations, and their thicknesses, is as follows:

AGE	FORMATION	THICKNESS
LATE EOCENE and OLIGOCENE		
	White River formation: *light clays and sandstones*	0–50 feet
PALEOCENE		
	Tongue River formation: *dark clays, sands, lignite*	0–300 feet
	Cannonball formation: *light clays and greenish shales*	0–150 feet
	Ludlow formation: *dark gray sands, clays, lignites*	0–200 feet

CRETACEOUS

Hell Creek formation:		
dark clays, sands, lignites	0–450 feet	
Fox Hills formation:		
light sands, greenish gray shales	0–200 feet	
Pierre formation:		
dark gray fissile shales	about 2,000 feet thick, but only upper few hundred feet are exposed	

The Pierre shale was deposited in the shallow waters of the inland sea, which withdrew from this area before the latest Cretaceous Hell Creek formation was laid down. The Hell Creek formation is famous for its dinosaurs, including the fearsome *Tyrannosaurus rex*. The dinosaurs vanished as the last of it was laid down. The inland sea again flooded the area, for the last time, as the Cannonball member was deposited.

Because the highway mostly follows the ridges, outcrops are poor except for the Pierre shale in the breaks of the Missouri and Grand Rivers. The strata are soft and weather to gently undulating prairie. Only a few layers within the soft sediments are resistant enough to make rock caps on buttes.

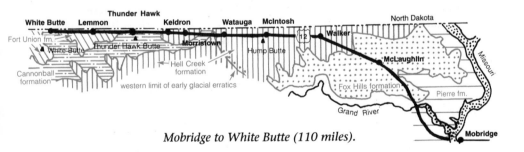

Mobridge to White Butte (110 miles).

Between Mobridge and McLaughlin

In early homesteading days, three railroads built westward routes across northeastern South Dakota, each hoping to be the first to bridge the Missouri River and tap the vast cattle and agriculture market to the west. The Chicago, Milwaukee, St. Paul, and Pacific won, completing a bridge across the Missouri River at Mobridge in

1906 and ultimately extending its rails to the Pacific coast. Mobridge is a contraction of "Mo. Bridge," as it appeared on early railroad construction reports. Except when the river was frozen over, vehicular traffic and cattle herds continued to use a ferry until the state completed a highway bridge in 1926.

The original bridge crossed north of the mouth of the Grand River. The creation of Lake Oahe made it necessary to route the highway south of the mouth of the Grand River and to build a new bridge there.

Grand River

The Grand River drains the northwestern edge of South Dakota. Before the ice ages, it flowed northeast across the eastern Dakotas, joined the Red River of the North, and finally emptied into Hudson Bay. Glacial ice advancing over the eastern Dakotas dammed the Grand River. With other similar streams, it finally found its way southward around the ice front and became part of the Missouri River system. Before the ice ages, the channel of the Grand River was 300 feet higher than at present. As the Missouri River entrenched its channel during the ice ages, all of its tributaries also eroded their channels to keep pace, and they extended their headwaters to the west.

West of the Grand River crossing, the highway climbs out of the breaks of the Missouri River and onto the Fox Hills upland. Sparsely vegetated hills give way to prairie densely covered with grass. A cemented stream gravel, correlated with the Bijou quartzite in the south-central part of the state, caps buttes on the skyline to the northeast.

McLaughlin is on the outcrop of the sandy Fox Hills formation, the principal shallow aquifer of the area. The municipal wells are 130 to 150 feet deep.

Between McLaughlin and White Butte

A few miles west of McLaughlin, the highway climbs above the Fox Hills sandstone and onto the clays of the Hell Creek formation. In general, the grass is poorer on the gumbo flats of the Hell Creek formation than on the Fox Hills sandstone.

McIntosh stands on the Ludlow formation, which was deposited during Paleocene time, not long after the dinosaurs vanished. The low hills 2 to 3 miles northwest of town are a remnant of the Cannonball formation. A bed of lignite coal at the southeast corner of the low hills provided a local source of fuel for many years. Wells at McIntosh produce water from the Fox Hills sandstone at a depth of 200 to 220 feet.

For several decades before World War I, northwestern South Dakota was open range, used by such famous cattle outfits as Turkey

Track, Hat, Matador, VVV, and Diamond A. With the coming of the homesteaders and barbed-wire fences, the big operators gave way to many smaller outfits. A few of the larger operators managed to continue until about World War II by leasing huge tracts of Indian land on the Standing Rock and Cheyenne River Sioux Reservations.

Bedrock here is clay deposited on dryland during late Cretaceous and Paleocene time, after the inland sea drained. It is not possible to differentiate the units from a car window.

Hump Butte is southwest of McIntosh, about 2.5 miles from the highway between McIntosh and Watauga. It is a remnant of the upper beds of the Ludlow formation that stands 150 feet above the prairie.

Widely scattered glacial boulders as far west as Watauga show that a very early glacial advance reached 40 miles west of the Missouri River. Erosion has removed the rest of the glacial deposits that once must have covered this area.

Bedrock is Ludlow formation; wells to the Fox Hills aquifer range from 290 to 330 feet in depth. Near Morristown, and again at Keldron,

Edmontosaurus regalis, a dinosaur collected from the Hell Creek formation near Watauga. —S.D. Museum of Geology

the highway dips into the valley of Hay Creek. Bedrock there is Hell Creek shale, but terraces along the stream consist of rubble eroded from the overlying Cannonball formation.

Thunder Hawk Butte, 2 miles to the southeast, also provides the name for the settlement. The butte is a remnant of the Cannonball formation that stands about 100 feet above the prairie.

Petrified Wood

Petrified wood is locally very abundant in the Tongue River formation. Most of it belongs to the redwood family. Logs reach a diameter of several feet, and stumps still in place measure up to 20 feet across the base. During the period 1930 to 1934, men hired with federal funds collected and hauled large chunks and logs to Lemmon, where stone masons created the Petrified Wood Park.

White Butte

The settlement takes its name from White Butte, a landmark 6 miles south of town. A bed of pale clay in the Chadron formation caps the 300-foot butte. It rests on dark gray Tongue River clays. Early settlers mined a seam of lignite coal near its base for domestic fuel.

Surveyors marked the boundary between the two Dakotas in 1891–92 using posts of Sioux quartzite set at half-mile intervals. The monuments were quarried in Sioux Falls, then transported by train, riverboat, and wagons to their final locations. Each is 8 to 10 inches square, 7 feet long, and set 3 feet into the ground.

White Butte, south of the town of White Butte. Light-colored clays and channel sandstones cap the butte.

The route between Fort Pierre and the area just east of Wall is within the drainage basin of the Bad River. The bedrock is all Pierre shale. For 10 to 15 miles west of the Missouri River, scattered erratic glacial boulders show that a very early ice advance once extended that far west. Stream gravels cap isolated buttes that stand above the prairie level. The gravel caps must be remnants of an old erosion surface, so the plains were once that much higher than now.

The Missouri River trench in the Pierre area is less than 3 miles wide and more than 300 feet deep. It was eroded when glacial ice blocked the streams that had flowed east, forcing water to find a new route southeastward along the ice front.

In the central part of the state, the Pierre shale is about 1,000 feet thick; only the middle section crops out in the breaks near Pierre. The siliceous DeGrey member tends to weather into stepped out-crops. The Verendrye member above it is mostly clay; it weathers into sticky gumbo. Both are best exposed along the Missouri and up the Bad River for several miles. Below are members of the Pierre for-mation in central South Dakota:

MEMBER	THICKNESS	DESCRIPTION
Elk Butte	160 feet	light gray shale; weathers to gumbo; only small remnants on highest ridges
Mobridge	0–100 feet	calcareous shale; weathers buff to yellow; found grassed over on high ridges
Virgin Creek	100 feet	upper: mudstone and clay; weathers to gumbo lower: shale; weathers to small flakes; numer-ous bentonite beds
Verendrye	160 feet	olive gray clay and mudstone
DeGrey	150 feet	light olive gray siliceous shale; many thin ben-tonites cause stair-step weathering; black manganese nodules
Crow Creek	10 feet	light gray marl with thin silty sand at base
Gregory	140 feet	gray claystone and shale
Sharon Springs	95 feet	black carbonaceous shale with many bento-nites; rests unconformably on Niobrara chalk; often with waterworn vertebrate fossils at contact

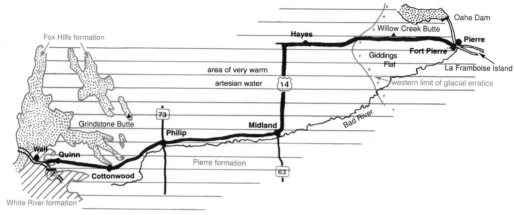

Pierre to Wall (114 miles).

When Lewis and Clark came this way in 1804, the Indians called the Bad River the Teton River; the white traders and trappers called it the Little Cheyenne River. Clark opted for the Indian name. Spring ice jams that caused frequent floods along the lower reaches finally inspired the local people to call it the Bad River. By whatever name, it drops 8 feet per mile between Philip and Midland, 5 feet per mile from Midland to its mouth at Fort Pierre. When the Chicago and Northwestern Railroad built west from Pierre to Rapid City in 1907, it followed the Bad River to Wall. US 14 crosses the uplands north of the river to Midland, thus avoiding the narrow bottoms of the lower Bad River and the frequent floods.

The highway quickly climbs to the uplands and crosses them along the entire 62 miles to Midland. Between Midland and Philip, the road follows the Bad River; between Philip and Wall, it follows the river's headwater tributaries for 55 miles. Slumps and small landslides within the Pierre shale explain the hummocky topography on steeper slopes.

The Verendrye brothers spent many months in the early 1740s exploring the upper Mississippi and Missouri Valleys. In 1793, they buried an inscribed lead plate intended to claim this vast region for France. In 1913, a group of young people found an encrusted plate of lead on a shale hill just north of Fort Pierre. It tallies in all details with the description of the one the Verendrye group buried. You can see the venerable relic in the South Dakota State Cultural Center in Pierre.

Willow Creek Butte

Willow Creek Butte, with its radio tower, stands 130 feet above the prairie. It was a landmark and stopping place for wagon trains

114

Undulatory topography is characteristic of landslides on the Pierre shale. —H. E. Simpson, U.S. Geological Survey

on the stage and freight route between Fort Pierre and the Black Hills. Pebbles in the gravels that cap the butte include pieces of limestone and chalcedony eroded from the White River Oligocene sediments to the west. They also include feldspar and quartz eroded from the pegmatites in the central Black Hills.

Fossil horse teeth from comparable gravels on other high buttes in the vicinity date the age of the deposit as older than the Illinoisan glacial stage. The gravel may record an old course of the Bad River. It now flows 500 feet below the level of the gravels on the butte, impressive testimony to the tremendous amount of erosion that has occurred in this area since middle Pleistocene time.

You can think of Willow Creek Butte as an example of inverted topography in which gravels deposited in a stream later protect the bedrock beneath them from erosion. Those gravels then become high points. The butte is on a ridge that apparently diverted the local advance of an early glacier. Glacial erratics are common on the northeast side of the ridge, but absent from its southwest side.

Township Wells

During the homesteading years before World War I, almost every parcel of 160 acres was a separate farm. The homesteaders could not afford to drill that many deep artesian wells, so the township trustees drilled community wells. Willow Creek Township had at least three. Most were in low areas to ensure a flow and to shelter them from the wind. Anyone in the community could haul water for stock or domestic use.

Fort Pierre Meteorite

The first meteorite identified in South Dakota was found 20 miles northwest of Fort Pierre in 1856. It weighed 35 pounds. Polished slices prepared in St. Louis showed that it was a nickel-iron meteorite. Those typically contain a little less than 10 percent nickel and significant amounts of other metals. That is a pretty good stainless steel, and it explains why nickel-iron meteorites weather so slowly. One slab of this one is in the St. Louis Academy of Science; others are in the American Museum of Natural Science and the British Museum.

Haakon County Geothermal Area

Between Hayes and Philip, the highway crosses a poorly defined area where artesian wells yield water that is abnormally warm for its depth. Warm water from a well drilled into the Cretaceous sandstones heats the Stroppel Hotel in Midland. Since the 1960s, wells to the Madison limestone have become sources of geothermal energy as well as municipal water supplies. Water from the Madison limestone heats school and county buildings in Midland and Philip. Warm water from a similar well west of Hayes heats farm buildings and dries grain. In some installations, warm water from flowing wells circulates through pipes embedded under the floors of buildings.

Philip

In 1961, the people of Philip became disenchanted with their municipal water supply, then a small reservoir 2 miles north of town. They drilled a well into the Madison limestone at a depth of 4,010 feet. The initial flow was 1,000 gallons per minute at a temperature of 153 degrees Fahrenheit, with a surface pressure of 240 pounds per square inch and about 1,100 parts per million of dissolved solids. They drilled next to Lake Waggoner so they could divert the water into the lake to cool it and to dilute the lake water, which contained even more dissolved solids. They treat the resulting blend in the original water plant at the lake before piping it to town. Warm well water circulated through simple heat exchangers heats the treatment plant.

Wall

Watch just east of Wall for the first remnants of the White River Oligocene clays lying on the Pierre shale. Wall perches on those clays on the Badlands Wall, a steep erosional scarp that extends south without a break for 9 miles. It separates drainage to the east to the Bad River from drainage to the west to the Cheyenne River.

Wall also decided in 1961 to switch to a deep well water system. Now, the town has several wells that produce from the Fall River and Lakota sandstones at a depth of 3,200 feet. The water rises to about 500 feet below the surface.

The big concrete dinosaur at Wall has no geological significance. The dinosaurs vanished at the end of Cretaceous time, about 35 million years before deposition of the White River Oligocene group on which Wall stands. No dinosaurs in that stuff.

US 18
Missouri River–Maverick Junction
288 miles

Between the Missouri River and the eastern edge of the Black Hills, US 18 crosses through many hundreds of vertical feet of pale sand, clay, and volcanic ash, some deposited by streams, some by the wind. All were laid down during and after the rise of the Black Hills. The sedimentary layers typically thicken to the west, toward their source.

In the western part of the state, you can follow many individual sequences of layers from outcrop to outcrop. Not so in the Missouri River area. The formations there are much thinner than those farther west, and some are locally missing. They are also finer and more homogeneous, so it is hard to divide them into smaller units.

During the rise of the Black Hills in Paleocene and early Eocene time, most of the debris of their erosion moved east beyond South Dakota, all the way to the Gulf of Mexico. The sedimentary formations deposited during Cretaceous time, first in the inland sea, later on land, were eroding as the Black Hills rose. During this period of warm and wet climate, the iron minerals in the dark shale weathered to shades of red, orange, and yellow.

Toward the close of Eocene time, the climate dried and the streams no longer had enough flow to carry their loads of sediment. Instead, they deposited that debris on the old land surface, widely burying the red soil that had weathered during Eocene time. Deposition in southwestern South Dakota was not continuous, but alternated with periods of erosion. These erosional breaks created irregular and undulating boundaries, unconformities, between successive sequences of sediments. In some places, deep valleys eroded, then filled in during the next period of deposition. Thus, we sometimes see younger sediments abutting valley walls of older rock.

117

Compaction and groundwater movement have greatly changed the original soft sand, clay, and ash beds. Calcium carbonate, dissolved in the water, precipitated, cementing the sediments. In some places, calcite precipitated in nodular concretions and ledges that stand out in relief on weathered outcrops.

After deposition of the last sands in late Miocene time, the cementing material was mostly quartz. In many cases, the cementation was incomplete, producing a weakly cemented sandstone. In many other cases, cementation filled all the voids, producing an absolutely solid rock in which both the sand grains and the cement between them are quartz. The resulting rock is quartzite, in which fractures break across sand grains, rather than around them.

Much of this quartzite is greenish gray. It is known as Bijou quartzite because it makes flat caps on the tops of the Bijou Hills, a group of buttes east of the Missouri River about 30 miles above the Fort Randall Dam. All occurrences of greenish quartzite are not exactly the same age, so the term Bijou refers to a rock type rather than a formation. The rough alignment of similar buttes east and west of the Missouri River suggests that the quartz cement may have precipitated from groundwater beneath large streams that flowed when the land surface was higher.

Some of the best range and farmland in western South Dakota is next to US 18. The high sand content of the underlying bedrock provides a loose soil, which absorbs and retains water better than the gumbo soils developed on the older Cretaceous clays and shales. Grass covers most of the area, making it hard to recognize the various rock formations. On newer highways, the highway department graded the roadcuts and planted grass on them, spoiling the view of the rocks.

Fort Randall

The army established Fort Randall in 1856, erecting 40 buildings made of cottonwood logs. In 1871–72, they replaced the original structures with frame buildings set on stone foundations. In 1875, the soldiers who belonged to the Odd Fellows Lodge built a stone chapel on their own initiative. The building was made of rough blocks of Niobrara chalk, faced with hand-sawn blocks of the same stone. When the army abandoned the post in 1892, it salvaged most of the wooden buildings, but the chapel remained, subject to vandalism and weathering. Recently, the building has been partially restored and designated a National Historic Landmark.

Old chapel. —S.D. State Archives

Between Fort Randall Dam and Martin

The lower bluffs along the river below the dam expose roughly 50 feet of Niobrara chalk. The main breaks expose the overlying dark gray Pierre shale. We often find waterworn bones of large swimming reptiles at the contact of the chalk with the overlying shale. Their waterworn character suggests that the sea probably retreated for a short time after it deposited the chalk, then flooded back to deposit the Pierre shale. If so, the two formations record two periods of flooding.

From the west end of the Fort Randall Dam, the road climbs through about 500 feet of dark gray marine shale of the Pierre formation. A remnant of fine sand of the Miocene Valentine formation caps the upland. If the White River beds had ever extended this far

Fort Randall Dam (Pickstown) to Winner (81 miles).

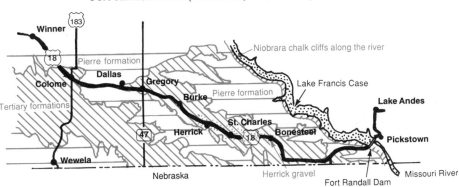

119

Period	Epoch	Group	Formation	Description	Thickness
	Recent		Sandhills formation	Dune sand, mostly stabilized with prairie grasses	
			— unconformity —		
Tertiary	Miocene	Ogallala group	Ash Hollow formation	Sandstone, fine grained, limy, with ledges of white limestone; hackberry seeds	0–150 feet
			Valentine formation	Sandstone and siltstone, poorly consolidated; cemented channel sands form Bijou quartzite	0–200 feet
			— unconformity —		
		Arikaree group	Rosebud formation	Clay and siltstone, pale pink to gray	0–300 feet
			— unconformity —		
			Harrison formation	Siltstone and sandstone with much volcanic ash; hard ledges give stair-step aspect to outcrops	0–160 feet
			— unconformity —		
			Monroe Creek formation	Sandstone and siltstone, tan to buff, wind deposited in part	0–250 feet
			Sharps formation	Siltstone and clay, pink, with layers of small concretions; Rocky Ford ash bed at base	0–35 feet
	Oligocene	White River group	Brule formation	Clay and siltstone, gray to pink, concretion zones; weathers to stair steps; channel sandstones contain minerals derived from Black Hills	0–400 feet
	Eoc.		Chadron formation	Clay, greenish gray, bentonitic, weathers to low hummocks with popcorn surface	30–70 feet
			— unconformity —		
Cretaceous			Pierre formation	Shale, dark gray, marine in origin	1,200–2,000 feet

Principal nonmarine rock units, southwestern South Dakota.

east, erosion removed them during middle Miocene time. The high-way follows the outcrop of sand to the eastern edge of Bonesteel, where it is overlain by much coarser sand and gravel of the early Pleistocene Herrick formation.

Between Bonesteel and Martin, the highway follows an old spur of the Chicago and Northwestern Railway, which followed the divide between Whetsone and Parma Creeks. The roadbed crosses sand and gravel of the Herrick formation to a point 4 miles west of Herrick. Between Herrick and Gregory, it crosses the fine sand of the Valentine formation. Between Gregory and Dallas, the road is on Pierre shale. Resistant remnants of Bijou quartzite within the Valentine formation cap buttes both north and south of the highway. All of the small towns along this stretch get municipal water supplies from shallow wells drilled into the Valentine sand or Herrick gravel. None of the many active and inactive gravel pits are visible from the highway.

Bedrock between Dallas and Winner is Pierre shale, except for a 4-mile expanse of sand in the Valentine formation from which Colome has developed a shallow supply of good water. The Colome Hills, southeast of Winner about 4 to 6 miles are Pierre shale capped with a resistant bed of Bijou sandstone and quartzite. Winner stands on the Pierre shale, which is not a good source of water. It developed a water supply from wells drilled into the Valentine sand 8 miles south of town.

Bedrock between Winner and Carter is Pierre shale. Gravel deposited during the ice ages caps some small hills a mile or so west of town; farther south Bijou quartzite caps some low buttes.

The Red Buttes just northeast of Carter are a northern remnant of the high tablelands that rise about 340 feet above the prairie in the area about 5 miles south of the highway. Bedrock at their base is Pierre shale, and above that is nearly 100 feet of the White River

Winner to Martin (103 miles).

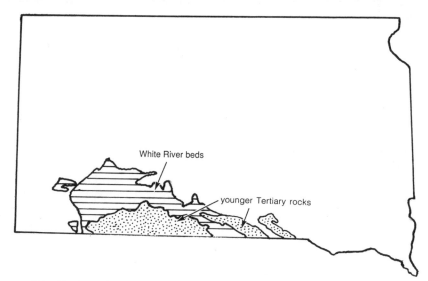

Distribution of nonmarine Tertiary rocks in southwestern South Dakota.

formation. Miocene Bijou quartzite caps the White River beds. The missing early Miocene strata were deposited and eroded before deposition of the late Miocene Valentine sand. Pits high on the butte excavate sand and quarry and crush the quartzite for road metal.

US 18 enters the Rosebud Sioux Indian Reservation just west of Carter. The road crosses Pierre shale between Carter and the area a mile east of Okreek, nestled in the valley of Oak Creek. Between Okreek and Mission, US 18 climbs through successively younger Tertiary rocks, but exposures of the Oligocene and Miocene rocks are very poor. A thin layer of resistant sandstone of the Miocene Ash

Continental sediments near Martin.

Recent	Sandhills formation		Sand, fine, gray, loose
Miocene	Valentine formation		Sand, fine, gray
	Harrison formation		Sand, fine, gray to olive, with concretions and ledges 150 feet
Oligocene	Monroe Creek formation		Sandstone, fine, pink to brown 250 feet
	Sharps formation		Siltstone, buff to pink, with Rocky Ford ash at base 150 feet
	White River group		Clays and siltstone, greenish gray to tan: concretions, ledges, and channel sandstones 300 feet
Cretaceous	Pierre formation		Dark gray marine shale 800 feet

Vegetation attempting to stabilize sand dunes south of Martin.

Hollow formation caps Haystack Butte, 6 miles east and 1 mile south of Mission.

The highway between Mission and Martin is on late Oligocene and Miocene sediments, except in a short interval where it drops onto White River clays in the valley of the South Fork of the White River. Most of the upland is underlain by the sandy Valentine formation; valleys are underlain by Miocene clays. From Vetal to Martin, the Miocene strata are at the surface, but outcrops are few.

From high points along the highway in the Martin area, the northern edge of the Nebraska Sandhills appears as a low ridge on the southern skyline. This vast sea of sand dunes is one of the largest tracts of dune sands in North America. They now lie still beneath a cover of grass. No one seems to know when they last marched before the wind.

Between Martin and Pine Ridge

Bedrock between Martin and Pine Ridge is mostly Miocene Rosebud formation, except where the road drops into the valley of White Clay Creek at Pine Ridge. Outcrops are few.

South of Denby, the Nebraska Sandhills extend nearly as far north as US 18. Several small buttes around Denby are made of soft Rosebud clay. Thin caps of the better-cemented Ash Hollow formation protect them from erosion. The highest bears the awesome name of Mount Denby.

Two miles west of Denby, a paved road heads north to Wounded Knee, site of the last armed conflict between Indians and the U.S. Army in 1891. Two miles northeast of Wounded Knee, but visible

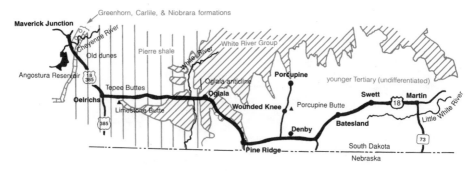

Martin to Maverick Junction (104 miles).

on the skyline from US 18, is Porcupine Butte. It is a forested mound of Rosebud clay capped by about 30 feet of the Ash Hollow formation. About 12 feet below the top is a conspicuous bed of white volcanic ash.

Pine Ridge is the administrative headquarters for the Oglala Sioux tribe. It takes its name from a prominent ridge that starts about 5 miles south of town and swings westward across northern Nebraska and eastern Wyoming for more than 100 miles. South of the ridge, the High Plains extend along the Rocky Mountain front to the Staked Plains of the Texas panhandle and beyond. North of the ridge, the country lies 400 to 600 feet lower.

Late Miocene and Pliocene sediments, largely sand and gravel, cover the High Plains south of Pine Ridge. Countless windmills and pumps that dot the upland surface produce water from shallow aquifers. They even furnish water for sprinkler irrigation. North of Pine Ridge, most of the late Miocene and Pliocene sand and gravel is missing, and the clays of the White River group and Pierre shale are widely exposed at the surface. They contain no aquifers worth mentioning.

White Clay Fault

A mile and a half south of Pine Ridge, SD 407 crosses a fault. It is inconspicuous at this point and cannot be traced far to the east. West of town, the fault trends northwest for many miles toward the Black Hills, but dies out before reaching the Cheyenne River. It is very conspicuous a short distance west of Pine Ridge, where the south side has risen far enough to bring Niobrara chalk into contact with pale gray Miocene strata. The amount of vertical displacement is not known. The fault breaks beds of Miocene age, so it is younger than them, therefore much younger than the stresses that raised the Black Hills in Paleocene time.

Gently dipping White River beds on an anticline north of Oglala. —J. E. Martin

Between Pine Ridge and Maverick Junction

The highway between Pine Ridge and Oglala follows the valley of White Clay Creek. Bluffs are outcrops of Sharps and Monroe Creek formations of late Oligocene age.

Look on the horizon a couple of miles north of Oglala to see a gentle arching of the White River strata, a broad anticline. Those rocks were laid down after the Black Hills rose, so the forces that bent them are much younger than those that raised the Black Hills.

Early oil exploration in South Dakota involved cable-tool drilling. A heavy chisel-like bit was dropped to break the rock into fine fragments, which were bailed from the hole.

Breaks of the White River west of Oglala. Developing badlands topography breaks the prairie sod.

Anticlines commonly trap oil, so the structure north of Oglala may be a potential oil field, but more study is required to justify the cost of drilling a test hole.

Bedrock between Oglala and Oelrichs alternates between pale clays of the White River group and the underlying dark gray clays of the Pierre formation. Oelrichs is on Pierre shale, probably 2,200 feet above the first sandstone artesian aquifer. Oelrichs developed a municipal water supply from wells less than 50 feet deep in the sandy alluvium of Horsehead Creek.

Fossil box turtle weathering from clay in the Brule formation, southwestern Shannon County. —Phil Bjork, S.D. Museum of Geology

Bedrock between Oelrichs and Maverick Junction is predominantly Cretaceous shale, mostly well covered with grass. The road cuts successively down through the geologic section, from the Pierre shale at the east end through the Niobrara, Carlile, and Greenhorn formations. Maverick Junction is on the Belle Fourche shale just below a rim of Greenhorn limestone.

On some of the high ground between Oelrichs and Maverick Junction, an irregular mantle of mostly stabilized dune sand overlies the shale. The wind picked up the sand from along the floodplain of the Cheyenne River, then dropped it away from the turbulence of the river breaks.

Limestone Butte

Two miles southeast of Oelrichs is an erosional remnant of the Chadron formation, part of the White River beds. It consists of about 135 feet of pale green and pink clay protected from more rapid erosion by three thin lenses of limestone in the upper 22 feet. The butte stands at the northwestern end of a remnant ridge of White River clay that extends nearly to the White River.

Tepee Buttes

Watch for numerous small pointed buttes north and south of the highway between Limestone Butte and Oelrichs. Their conical outlines inspired the name tepee buttes, which is in use along hundreds of miles of the high western plains. Similar buttes exist all around the Black Hills, invariably at two distinct levels within the Pierre shale. Their summits bear a limestone cap, which protects the soft clay beneath from erosion. Small clamshells cram the limestone. Some geologists think the patches of limestone that cap the tepee buttes were deposited around warm springs on the seafloor.

US 83
Nebraska Line–Pierre
122 miles

The southern third of this route crosses pale sands and clays deposited on dryland during late Oligocene time. North of the area about halfway between Mission and White River, the bedrock is dark Pierre shale. It was deposited in shallow seawater during late Cretaceous time.

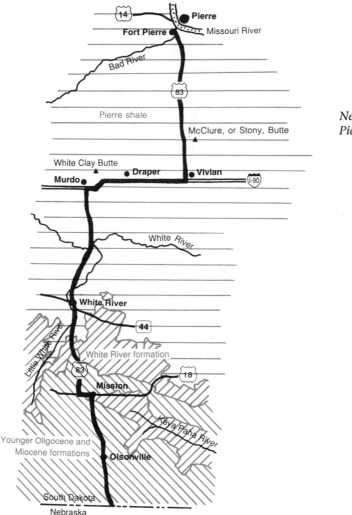

14

● Pierre

Fort Pierre ●

Missouri River

Bad River

83

Pierre shale

McClure, or Stony, Butte ▲

*Nebraska line to
Pierre (122 miles).*

White Clay Butte ▲

Murdo ●   ● Draper   ● Vivian

I-90

White River

White River

44

White River formation

83   18

Mission

Keya Paha River

Younger Oligocene and
Miocene formations

Olsonville

South Dakota
Nebraska

Bedrock between the Nebraska state line and Pierre is the Sharps and overlying Monroe Creek formations, both of late Oligocene age. Sluggish streams originating far to the west deposited these beds on a broad alluvial plain. Part of the sediment eroded from older clays; much of it was volcanic ash that erupted from volcanoes near the west coast and drifted east on the wind. Notice the persistent grain of the landscape, with all streams flowing to the southeast.

Mission is on the north side of Antelope Creek, whose valley has locally cut down through the younger strata and exposed the underlying White River beds. Mission secures its domestic water supply from wells less than 200 feet deep into the Arikaree formation.

North of US 18, the highway descends into the breaks of the South Fork, or Little White River. The river cut down through late Oligocene beds across a belt several miles wide, went through the underlying clays of the White River formation, and dug into the somber gray marine shales of the Pierre formation. The town of White River gets its water supply from very shallow wells in the alluvial gravels along the South Fork of White River.

The highway between White River and Murdo roughly follows the breaks of the South Fork of the White River. Both White Rivers owe their names to the quantities of pale clay in suspension in the water. North of the White River, the road gradually climbs to Murdo on the divide between the White and Bad Rivers. The town had to drill 3,300 feet to find an aquifer with potable water. Watch about 3 miles northeast of Murdo for White Clay Butte with its cap of pale clays and gravel, probably a remnant of the White River formation. Draper gets its water supply from two wells drilled 2,200 feet into the Dakota sandstone.

Nine miles north of Vivian watch for Stony, or McClure, Butte, also capped with pale rock, probably another remnant of the White River formation. The upland surface between Vivian and Pierre contains many shallow depressions, blowouts eroded by the prevailing northwest wind. Some hold water in wet years. The descent into the Missouri River Valley at the mouth of Bad River is abrupt, about 400 feet in 5 miles.

US 85
Spearfish–North Dakota Line
107 miles

Going northward, the route crosses progressively younger rocks. They range in age from Permian and Triassic redbeds at Spearfish to buttes capped with Miocene sediments near the North Dakota line. They represent about 325 million years of geologic time. Most of the sedimentary formations along the southern half of the route were deposited in shallow seawater. Those along the northern part were deposited on a broad alluvial plain, probably in a very dry climate.

129

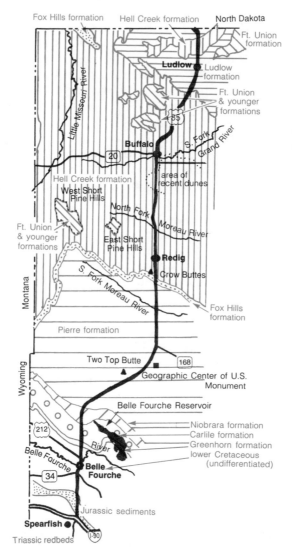

Fox Hills formation   Hell Creek formation   North Dakota

*Spearfish to North
Dakota line (107 miles).*

Those deposits are mostly soft shales and clays that weather to a gently undulating prairie surface. Some hard layers in the younger rocks permit the local development of buttes and tablelands.

The route also offers a glimpse of the many facets of the economy of the state. In the Black Hills, tourism, mining, and lumbering dominate the economy; north of the Black Hills, sheep and cattle ranching form the mainstay; and in the Buffalo area, a modest oil industry adds to the ranching income.

Between Spearfish and Redig

The highway crosses redbeds of the Spearfish formation as far north as the Redwater River. They were deposited during Permian and Triassic time, a period when red sediments accumulated in many parts of the world. After storms, the river may run red for several days because it carries so much sediment eroded from the large area of Spearfish redbeds to the west.

North of the Redwater River, the route parallels the west flank of the gently folded Belle Fourche anticline. The layers of sandstone that tilt down to the west along the eastern side of the road were deposited during late Jurassic and early Cretaceous time. Before it was rebuilt, sections of this road had a tendency to slide down that dip into the ravine.

Thick beds of clay deposited on land are sandwiched within the layers of marine sandstone. Between 1927 and 1978, a mine quarried these clays and shipped them to Belle Fourche to make brick and tile. Since the brickyard closed, most of the brick used in this area has come from Hebron, North Dakota. The hills with pine trees on them west of the highway and south of the junction with SD 34 are the outcrop of siliceous Mowry shale.

Belle Fourche is at the junction of the Redwater and Belle Fourche Rivers. A large meander loop of the Belle Fourche River embraces the rodeo grounds.

In the 14 miles between the north side of the Belle Fourche River and Owl Creek, the highway crosses a succession of low ridges held up by ledges of limestone in the Greenhorn, Carlile, and Niobrara formations, all deposited during Cretaceous time. Some contain abundant concretions. The airport is on a ridge underlain by the Greenhorn limestone. Bedrock between Owl Creek and the Moreau River is the dark Pierre shale.

Two Top Butte

US 85 alters course to avoid a complex of high buttes west of the road. Two Top Butte is a remnant of Pierre shale capped by a bed of Fox Hills sandstone perched 500 feet above the highway. Originally, several hundred feet of late Cretaceous and Paleocene clays and sands were on the Fox Hills sandstone. Erosion largely removed them during Eocene time, when the climate was very wet. Then the climate dried and streams flowing from the northern Rocky Mountains and the Black Hills laid down the pale clays and sands of the White River group. Most of those beds were eroded during middle Miocene time, when the streams again ran full.

At Two Top Butte, you can see huge blocks of White River conglomerate strewn over the top and flanks of the hill. Erosion of a

	Recent		unconformity	Local areas of dune sand
	Miocene		Arikaree formation	Sand, fine and white, with much volcanic ash; well cemented; cliff former 0–260 feet
			unconformity	
Tertiary	Oligocene	White River	Brule member	Clays and sands, light gray to pink, with some channel sandstones and numerous concretions 0–440 feet
			Chadron member	Clay, greenish gray, bentonitic 0–50 feet
			unconformity	
	Eocene		Golden Valley formation	Clays, dazzling white to golden brown 0–70 feet
			unconformity	
	Paleocene	Fort Union fm.	Tongue River member	Clay, sandy, gray, and massive cliff-forming sandstones 0–300 feet
			Ludlow member	Cannonball member: sandstone and shale with round concretions, marine invertebrate fossils 0–225 feet
				Clays and sandstones, gray, interlayered with beds of lignite coal 0–450 feet
Mesozoic	Cretaceous		Hell Creek formation	Clays, somber gray, with channel sandstones and some thin beds of lignitic coal; dinosaur bones 0–450 feet
			Fox Hills formation	Shoreline sands and lagoonal clays left by retreating Cretaceous sea 0–200 feet

Cretaceous to Miocene formations in northwestern South Dakota.

maximum of several hundred feet of the soft underlying clays slowly let those blocks down. Many of the blocks still move as the ground alternately freezes and thaws seasonally. Similar remnants cap other buttes aligned in an east-to-west direction, including Castle Rock, Deers Ears, and Mud Butte to the east.

Old stream channels in the White River group east of the Black Hills contain minerals and rocks clearly identifiable with a source in the central Black Hills, but these northern channels contain no material of Black Hills origin. The next nearest possible source is the Bighorn Mountains of Wyoming. So far, no one has looked to see if they were indeed the source.

Stone Monuments

The stone cairns, or stone johnnies, that top many buttes are a familiar site throughout the western states. Some originally marked wagon and cattle trails or indicated sources of water. Surveyors built others as triangulation stations for the first precise surveys of the area. But sheepherders built most of the monuments, particularly those on lesser buttes. They stacked rocks to idle away the hours while they watched their flocks graze. The distance a man can lift a slab of sandstone limits the height of a stone cairn, so most are only 5 to 6 feet high and 3 feet across. For a few days each season, when

Two Top Butte. The two peaks are capped with conglomerate formed from stream gravel of White River age. These "survivors" have been let down by erosion of the soft shale beneath them.

the harvester ants are swarming, myriads of flying and crawling insects swarm over those high points. I have seen my surveying instrument disappear in a seething ball of insects in only a few seconds as a flying swarm settled upon it, and on me.

Castle Rock Buttes

The isolated hills on the skyline several miles northeast of the junction with SD 168 are the Castle Rock Buttes. The highest and most southerly is Castle Rock, which stands nearly 600 feet above the prairie. The bedrock in this area is Pierre shale, which rarely erodes into prominent hills. The Castle Rock Buttes are islands of Fox Hills sandstone, which better resists erosion. Castle Rock has a cap of Fox Hills sandstone overlain by a remnant of Hell Creek clays and sandstones. Four miles farther north is Haystack Butte, or Square Top, which has a cap of pale clays of the White River beds.

A Splendid Dinosaur

Somewhere north of where the highway crosses the Moreau River, J. L. Wortman, working for the American Museum of Natural History in 1881, found and collected one of the most perfectly preserved dinosaur skeletons ever recovered. He named the specimen

Castle Rock Buttes.

Atlantosaurus copei to honor E. D. Cope, a noted vertebrate paleontologist of the time. The beautifully prepared and mounted specimen is still on display in the museum in New York City.

Crow Buttes

Several remnants of Hell Creek clays stand as much as 275 feet above the prairie west of US 85, just north of the Harding County line. In 1822, a war party of Sioux surprised a much smaller group of wandering Crow Indians. After a skirmish in which the Sioux captured the Crow women and children, the Crow warriors escaped to

One of the Crow Buttes.

Segment of cemented stream channel gravel, let down several hundred feet by erosion of the softer shales and clays. Near Center of Nations Monument.

the Crow Buttes, where they set up defensive positions. According to Indian tradition, the Sioux besieged the Crows and did not leave until they died of thirst.

On the east side of the highway, opposite the Crow Buttes, are huge blocks of sandstone and conglomerate resting randomly on the Hell Creek shale. The material is a channel sandstone, part of the White River sediments, perhaps a tributary to the major channel at Two Top Butte. But the sandstone blocks rest on the lower part of the Hell Creek shale. No one knows how far erosion of the higher section of the Hell Creek formation has let them down.

Between Redig and Buffalo

Redig was probably a stopping place along the stage route between Bowman and the Black Hills. It is a survivor among the many local post offices established during the homesteading days just before World War I. This is sheep and cattle country, with an occasional venture into dry-land wheat farming.

About 5 miles south of Redig, 1 mile south of the line between Harding and Butte Counties, the highway crosses a narrow and inconspicuous outcrop of the Fox Hills formation. These are sands and clays deposited along the shore as the inland sea retreated during late Cretaceous time. They also mark the boundary between the marine clays of the Pierre formation to the south and the overlying Hell Creek formation.

The Hell Creek strata appear to have been deposited in a low and poorly drained environment. The beds owe their somber color to

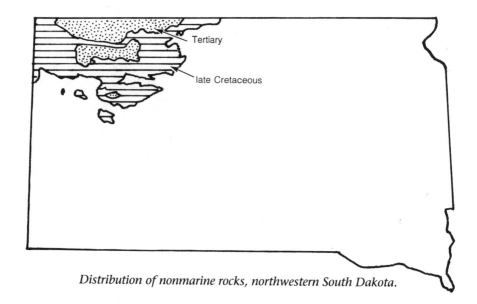

Distribution of nonmarine rocks, northwestern South Dakota.

organic material, and they exhibit such sedimentary features as mud cracks, raindrop impressions, ripple marks, and occasional tracks of birds and small animals. A good many excellent dinosaurs have come out of the Hell Creek formation, so we can imagine them swarming over those drying muds.

Buffalo gets its water supply from a series of wells drilled 90 to 130 feet into a sandstone aquifer within the Hell Creek formation. Over a wide area 2 to 8 miles south of Buffalo, the highway crosses a patch of recent, but well stabilized, sand dunes. The fine, dark sand comes from a sandy horizon near the middle of the Hell Creek formation, which crops out to the west, upwind.

Short Pine Hills

East and West Short Pine Hills stand 350 to 500 feet above the surrounding prairie several miles west of the highway. They are erosional remnants. Lignite coal beds in the Tongue River formation are at the base of each. Above the Tongue River formation is a section of pale clays and sands in the White River formation. The rimrock on top is massive sandstone, presumably deposited during Miocene time.

The Custer expedition visited the Short Pine Hills in 1874 on the way from Bismarck to the Black Hills. G. B. Grinnell, the expedition's naturalist, measured and described several outcrops of lignite coal.

The Jumpoff, southwest of Buffalo. The upland is capped with light clays and sands of the White River formation.

In 1977, oil geologists working in the West Short Pine Hills area discovered natural gas in the upper Cretaceous Shannon sandstone at depths of less than 2,000 feet.

Jumpoff Country

Four or five miles north of the East Short Pine Hills is a prominent escarpment that trends northwest. The rolling plateau to the south has an elevation between 3,400 and 3,500 feet. North of the escarpment lie intricately dissected badlands. The local name, the Jumpoff, applies both to the escarpment and to the badlands belt north of the escarpment. The badlands topography was carved from the Hell Creek formation, drab clays and sands deposited during late Cretaceous time and the source of many splendid dinosaur fossils. Pale sands and clays of the White River formation are conspicuous above the somber clays of the Hell Creek formation.

Cave Hills

Two large tablelands that stand 300 to 500 feet above the prairie southwest of Ludlow are the North and South Cave Hills. Like other tablelands in this area, they are remnants of the formerly continuous strata that blanketed the region during early Tertiary time. The lower part of the buttes is the Ludlow formation; the upper part is the Tongue River formation. The rimrock on both is a massive brown sandstone that makes them look like fortresses.

The name Cave Hills comes from an enlarged crevice eroded in the upper Tongue River sandstone 4 miles southwest of Ludlow. Captain William Ludlow, an army engineer assigned to the Custer expedition of 1874, visited this rude cave, already known to the Indians. His field notes for July 25, 1874, state that "the cave is a hole washed out of the sandstone 200 to 300 feet deep horizontally, with an entrance 15 or 20 feet, and proved to have no special interest other than that imparted to it by the superstition of the Indians." Archeological investigation of the cave yielded some trade beads and other relatively recent artifacts.

Lignite coal beds in the upper portion of the Ludlow formation locally reach a thickness of more than 10 feet. Other lignite beds near the middle of the Tongue River formation are mostly less than 3 feet thick. Many small "wagon mines" produced from these lig-

Entrance to Ludlow Cave.

The resistant Tongue River sandstone member of the Fort Union group caps many of the buttes in the Cave Hills. The pitted weathering is characteristic.
—C. J. Hares, U.S. Geological Survey

nites during the early homesteading days before World War I; a few reopened during the depression years of the 1930s.

During the late 1950s, prospectors discovered uranium in some of the lignite beds. They staked mining claims everywhere, but the small volume of the ore and the difficulty of separating the uranium from the lignite prevented much mining. The volcanic ash beds in the overlying White River formation probably provided the original source for the uranium. Groundwater, percolating downward, leached the uranium from the ash and redeposited it when it reached the chemical environment of a coal bed. Later erosion of several hundred feet of sediments exposed the coal at the surface.

Other Tongue River coals crop out in a belt that extends east from Ludlow to Ralph and Lodgepole, in Harding and Perkins Counties. Many small wagon mines operated during homesteading days. Mines worked seams that ranged from 6 to 15 feet thick.

Buffalo Oil Field

In the winter of 1953–54, an oil test drilled on the northwest flank of the Cave Hills, 15 miles northwest of Buffalo, came in producing from the Ordovician Red River dolomite at a depth of 9,529 feet. Initial production was 8 barrels of oil and 200 barrels of water per day. As it finally developed, the Buffalo field is about 9 miles long and 6 miles wide. Many smaller fields, developed later in Harding County, also produce from the Red River dolomite.

Missouri River–Wyoming Line

The long drive between the Missouri River and the Wyoming line crosses a rather uneventful landscape eroded mostly in sedimentary formations deposited during late Cretaceous time, some in the shallow inland sea, others on land slightly above sea level. Despite the general monotony of the landscape and the rocks, the road passes a number of geologically interesting features.

Missouri River Bridge

The agency for the Cheyenne River Sioux was on the west side of the river back when riverboats moved the freight. Forest City grew up across the river. A ferry connected them, except when the river froze over and people and cattle crossed on the ice. In 1924, the state built a bridge across the Missouri about a mile north of the present bridge. Whitlock's Crossing rose on the east bank during construction, and that was the end of Forest City. When the Oahe Dam flooded all three locations, the engineers moved the bridge to its present site, and the agency moved to Eagle Butte in 1959. The present bridge connects the sites of the two original settlements.

Between the Missouri River and Red Elm

The road crosses Pierre shale in the 51 miles between the Missouri River and Parade. Occasional erratic boulders scattered over the country east of La Plant record an early glaciation.

The road climbs out of the Missouri River breaks across successive units of the Pierre shale. Many landslides within the weak shale created the hummocky roadside topography. The full thickness of the Pierre shale in this area is at least 800 feet, some of which is below river level.

Missouri River to Faith (96 miles).

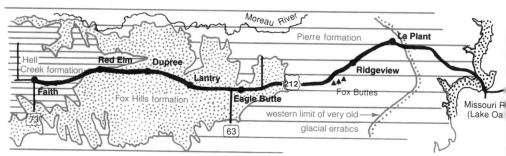

West abutment of the Forest City bridge across the Missouri River. The rise of reservoir water behind the Oahe Dam destabilized an ancient landslide in the Pierre shale, causing it to move. The huge area of lumpy topography marks the slide. —S.D. Department of Transportation

You can easily distinguish subunits within the Pierre shale. In some, the shale is dark gray and bentonitic and weathers to gumbo; in others, it is light gray and siliceous and weathers to silvery flakes. The gumbo sections are typically grassed over; the more siliceous units tend to outcrop in steps that are barren in dry years. In wet years, the grass makes it much harder to recognize the different units. Some contain limy concretions 2 inches to 2 feet across.

The highway between Parade and Red Elm parallels a railroad spur built to serve homesteaders in the days before World War I when every parcel of 160 acres sported its claim shack. The highway crosses a wide, but poorly defined, outcrop of Fox Hills sandstones and shales, which erode to a gently undulating surface. And they develop a loose, sandy soil that supports range grasses much better than either the overlying Hell Creek formation or the underlying Pierre shale. Trees planted as windbreaks on the Fox Hills sandstone are healthy, and the draws contain deciduous brush and trees. Sandstone aquifers in the Fox Hills formation furnish enough water for domestic use at depths of less than 250 feet.

The town of Eagle Butte was a small ranch settlement before the coming of the Cheyenne River Sioux Agency in 1959. In 1956, expecting the possible move and hoping to get better water than that from the 2,000-foot artesian wells into the Cretaceous sandstones, the town drilled a well 4,325 feet into the Madison limestone. It flowed more than 100 gallons per minute of hot water, 127 degrees Fahrenheit, and was capable of pumping much more. However, the water contained 2,347 parts per million dissolved solids, twice the maximum recommended by the U.S. Public Health Service. After establishment of the agency, the federal government built a pipeline to the Missouri River, and now the town has an adequate supply of properly treated water with a reasonable level of dissolved solids.

In 1963, the people of Dupree, wanting to supplement their shallow wells, drilled a deep well into a sandstone aquifer in the Fox Hills formation. It penetrated the Madison limestone between 4,268 and 4,500 feet. The well flowed water at 140 degrees Fahrenheit, but the dissolved mineral content was 2,300 parts per million, far too much.

Lantry Oil Field

In 1965, a wildcat oil test 4 miles north of Lantry discovered heavy crude oil at a depth of about 5,000 feet in the Ordovician Red River dolomite. That is the same formation that produces oil in several fields in the northwestern corner of the state. The first well failed to produce oil in commercial quantities. In the early 1970s, a private investment group randomly drilled an additional 18 holes around Lantry, mostly northwest of town. They developed modest commercial production.

Between Red Elm and Newell

The highway between Red Elm and Mud Butte crosses somber, lignitic shales of the Hell Creek formation, except in a few miles just east of Maurine.

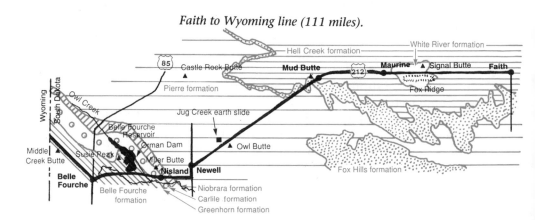

Faith to Wyoming line (111 miles).

Collecting dinosaur bones from the Hell Creek formation northwest of Mud Butte.

Fox Ridge rises 250 to 300 feet above the prairie about 2 miles south of the highway and between 2 and 5 miles east of Maurine. It is a narrow ridge that trends parallel to the highway as it divides the Moreau River drainage to the north from Sulfur Creek, a major tributary of the Cheyenne River, to the south. Fox Ridge is an erosional remnant of dark clays in the Paleocene Ludlow formation with a narrow crest of pale sands and clays of the White River group. The thin layer of limestone that protects the flat top from erosion contains cracks filled with moss agate. Moss agate is a variety of translucent chalcedony with small inclusions of iron or manganese oxide. Some inclusions form patterns resembling plants or landscapes, and are prized by makers of rock jewelry.

Fox Ridge is an erosional remnant of the formerly continuous blanket of White River sediments that once covered this entire region. Similar remnants include Signal Butte, north of the road east of Fox Ridge, the Slim Buttes farther north, the Big Badlands to the south, and many scattered buttes to the west.

In 1877, Ben Ash was deputy U.S. marshal out of Bismarck. He volunteered to use two wagons to transport a pair of federal prisoners to Deadwood and then return by the same route. The trail he pioneered soon became the route of important stage and freight traffic between the railhead at Bismarck and the new town of Deadwood in the Black Hills. A red granite marker sits where US 212 crosses traces of the old trail. A drive up the turnout will enable you to view the prairie for many miles to the east, south, and west.

Pierre shale landsliding into a cut bank.

Mud Butte Post Office gets its name from the only landmark in sight, an isolated erosional remnant of dark gray clay of the latest Cretaceous Hell Creek formation.

For several miles southwest of Mud Butte, US 212 crosses and re-crosses a narrow outcrop of Fox Hills formation, but the contacts are hard to see. The remainder of the route into Newell crosses gently rolling topography eroded on the Cretaceous Pierre shale. Deers Ears Butte is an erosional remnant of Hell Creek clays with two tops. Its cap is resistant sandstone and conglomerate in the lower layers of the White River group. Owl Butte, a mile southeast of the highway, is another remnant of Pierre shale; it stands 225 feet above the prairie.

About 10 miles northeast of Newell, the highway crosses Jug Creek, a small unmarked stream. Especially if you are driving toward Newell, watch for the beautiful set of slumps in the valley wall, just west of the highway. They developed as Jug Creek undercut a long slope in Pierre shale, which tends to slump anyway.

About 5 miles northeast of Newell, the road crosses a wide belt of low knobs called tepee buttes because of their general conical pro-file. Each is capped by a mass of limestone that resists erosion. The

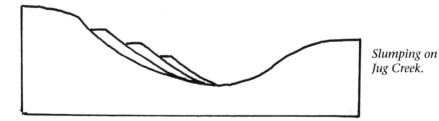

Slumping on Jug Creek.

limestone contains abundant fossils of small clams known as *Lucina occidentalis* and quite an assortment of small snails. These limestone patch reefs lie in a zone 900 to 1,000 feet above the base of the Pierre shale. They provide a wonderful glimpse of life in the shallow inland sea that flooded most of South Dakota during late Cretaceous time.

Between Newell and the Wyoming Line

The 25-mile route between Newell and Belle Fourche lies mostly in the alluvial floodplains of the Belle Fourche River and its tributaries. Bedrock outcrops are few and poor. Miller Butte, just north of Nisland, is an outcrop of late Cretaceous Niobrara chalk protruding through the alluvium of Owl Creek. A remnant of Greenhorn limestone caps Susie Peak, on the west side of Orman Lake.

Our route crosses the Belle Fourche Irrigation District, one of the oldest in the United States. In 1910, completion of Orman Dam across nearby Owl Creek diverted part of the flow of the Belle Fourche River through a 6.5-mile canal to supply the Belle Fourche Reservoir, also called Orman Lake. The dam is an earthen structure with a crest length of 6,200 feet and a height of 115 feet. Blocks of limestone riprap on the upstream face come from the Black Hills. When completed, it was the largest earthen dam in the world. Most of the impoundment is underlain by the dense Belle Fourche shale, but outcrops of Greenhorn limestone appear near the dam, and the upper reaches of the reservoir rest on Carlile shale. Collectors find many fine fossil shark teeth in that formation when the water is low. The

Shark teeth found in the Carlile shale at Orman Dam. —S.D. Museum of Geology

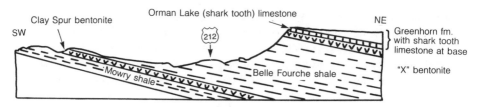

Section through Middle Creek Valley, northwest of Belle Fourche.

conspicuous northward swing of the Greenhorn limestone outcrop at the reservoir expresses the Whitewood anticline, a gently arching fold in the rocks.

The Harmon well is on the west side of the lake, about 1.5 miles north of the highway. It was drilled as a wildcat oil well, but it found artesian water instead. It produces several hundred gallons per minute from aquifers in the Minnelusa and Madison formations.

Belle Fourche started in about 1890, when the Chicago and Northwestern Railroad reached the area. It is an important trading center, a shipping point for cattle, wool, and bentonite clay. A sugar factory east of town processed sugar beets from 1927 to 1965, and a brickyard produced brick and tile products from 1927 to 1979.

In the 14 miles between Belle Fourche and the Wyoming line, US 212 follows the valley of Middle Creek, eroded in the soft Belle Fourche shale. The thin Orman Lake limestone member holds up the steep north side of the valley. Some people call it the shark tooth limestone because it contains so many of them. Middle Creek Butte south of US 212 is another remnant of Belle Fourche shale capped by the very thin but resistant Orman Lake limestone.

Ten feet below the Orman Lake limestone is a layer of bentonite about a foot thick, which petroleum geologists call the "X" bentonite bed. Thin as it is, that layer is traceable for hundreds of miles in oil and water wells. This bentonite started out as volcanic ash that almost certainly erupted in western Montana, where late Cretaceous time brought intense volcanic activity. The plume of ash drifted east on the wind and settled across the inland sea. Now it provides an absolutely precise time line for geologists who study the sedimentary formations laid down during late Cretaceous time. That is an enormous help in sorting out the layers of rock.

The low ridge with trees on it that parallels the highway a couple of miles to the south is the outcrop of the siliceous Mowry shale. At the top of the Mowry shale, and exposed on the long north facing dip slopes, is the commercially important Clay Spur bentonite bed.

The Bentonite Business

Bentonite is a smooth, yellowish or white clay formed by the chemical decomposition of volcanic ash. Most commercial bentonite deposits developed from ash that fell into a quiet sea, settled to the bottom without mixing with other sediments, and later was buried under other sediments. Given time, the ash reacts with seawater to form a clay mineral called montmorillonite. Most people rarely think much of clay and do not realize that the world of clay minerals is a universe in itself. All are variations on the theme of silicate layers stacked like pieces of paper. Montmorillonite is one of the kinds that swell like an expanding accordion as they absorb water and other substances between their silicate layers.

Hundreds of bentonite beds exist in the late Cretaceous shales of the northern High Plains. Each fell from a plume of volcanic ash drifting east from an eruption, probably in the northern Rocky Mountains. Picture a volcanic eruption large enough to spread a layer of ash over a dozen states, a layer 2 inches thick 800 miles from the volcano. The Clay Spur bed is 3 to 5 feet thick. Try to imagine the volcanic explosion that dropped that layer of ash all across the northern Black Hills and nearby Wyoming and Montana. The 1980 eruption of Mount St. Helens was a trifling event by volcanic standards!

Bentonites that contain a lot of sodium swell and absorb the most. If you cover the bottom of an excavation for a farm pond with sodium bentonite and then add water, the clay will swell into a watertight seal. Sodium bentonites are much sought after for use in sealing dams and canals. They are also used to make thixotropic drilling muds that flow easily, but set to a gel when they are no longer in motion. That is important because the flow of drilling mud sweeps the rock cuttings out of the hole. If drilling stops, the mud will gel and catch the cuttings, preventing them from settling to the bottom of the hole and locking the drilling tools.

Calcium bentonites do not swell, but they are useful as binders for such tasks as pelletizing iron ore, and as fillers in glossy paper. They even find their way into some foods, helping to make chocolates that melt in your mouth, but not in your hand.

The bentonite industry started in western South Dakota during the 1920s. Mining is a matter of stripping the overburden, exposing the clay to the weather for a year, then hauling it to stockpiles at the mill. Companies may blend clay from different pits to produce a uniform product. Milling consists of drying the clay in a rotary kiln, then pulverizing and sizing it to the consistency of flour. The milled clay is either sacked in 50- and 100-pound paper bags or shipped in bulk in special railroad cars. Distribution is worldwide. An important specialty product at Belle Fourche is absorbent kitty litter made from expanding montmorillonite.

Three bentonite processing mills were once active in Belle Fourche, but with depletion of the nearby deposits, mining shifted westward. One plant remains in Belle Fourche; the others have moved to northeastern Wyoming.

Boundary Markers

Large posts of pink Sioux quartzite set at mile intervals mark South Dakota's western boundary. The markers were quarried in eastern South Dakota in 1904, shipped by rail as far as possible, then hauled by wagon to their final site. The columns range from 10 to 12 inches across and up to 7 feet long. A typical stone stands south along the state line fence, about 1,100 feet south of US 212. Inscriptions on the four faces are "SD," "WYO," "1904," and "125." The last number is the distance in miles from the point common to South Dakota, Nebraska, and Wyoming.

Bentonite mill at Belle Fourche. Clay from stockpiles is blended, dried, pulverized, processed, and prepared for shipment worldwide.

SD 20
Mobridge–Montana Line
193 miles

SD 20 follows the broad divide between the Grand River to the north and the Moreau River to the south. Except in the Missouri River breaks at the eastern end and at the Slim Buttes near the western end, the soft upper Cretaceous and lower Tertiary continental sedimentary rocks erode to a gently undulating plain. They contain few resistant layers that might make cliffs, buttes, or box canyons.

Some of the rock formations weather to better soil than others, but they all lack hard ledges that form outcrops. Along this stretch of highway, we must rely upon the geologic map to tell us what formation lies beneath the prairie sod.

Before the homesteading era started in western South Dakota in about 1910, a few large cattle outfits ran thousands of head on the open range. The homesteaders brought wire fences and established numerous small communities. Most of these settlements are now gone, leaving one survivor every 20 to 30 miles along the main highways.

Between the Missouri River and Meadow

The long route successively crosses outcrops of Pierre shale, which was deposited in the Cretaceous inland sea, and the Hell Creek clays and sands, which were laid down on dryland after the land had risen above sea level. On top of the Hell Creek formation are the clays of the Ludlow formation, dark with lignite coal. They were deposited in Paleocene time, shortly after the dinosaurs vanished.

The Missouri River breaks west of Mobridge expose a thick section of Pierre shale—about 500 feet of it. The subdivisions of the formation are hard to recognize, especially in a wet year when the grass is lush.

Missouri River to Meadow (101 miles).

149

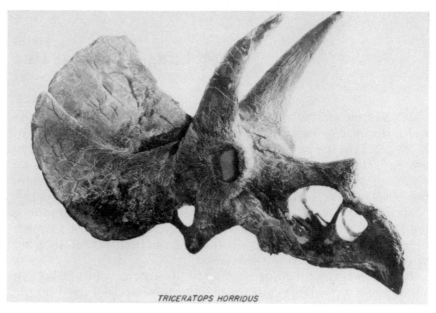

TRICERATOPS HORRIDUS

Triceratops skull 5.5 feet long, from Hell Creek beds. —S.D. Museum of Geology

The sandy Fox Hills formation caps the upland a few miles back from the river. Scattered large erratic boulders on the surface are all that remains to tell of a very early ice sheet that reached west of the present Missouri River nearly to Timber Lake. Most of these rocks are Precambrian granite and gneiss carried in the ice from many hundreds of miles to the northeast from the general region of the Hudson Bay.

Trail City is on the Fox Hills sandstone. Before the bridge was built at Mobridge, cattle were driven to Trail City from all over the northwestern quarter of the state. There they waited to ferry across the river in summer, or to cross on the ice in winter. Trail City has no municipal water system. Some of the private wells produce water from sands within the Fox Hills formation at a depth of less than 100 feet.

Timber Lake takes its name from the adjacent shallow lake, which floods a large blowout, a depression eroded by the prevailing north-westerly winds. The wind must have done that when the climate was much drier than now, perhaps during one of the interglacial periods. The lake depends entirely upon rainfall, so it changes in size with the seasons and with longer periods of wet or dry weather. The town of Timber Lake originally relied upon its lake for water, as did the railroad. Now, shallow wells into the Fox Hills formation supply a more reliable source of better domestic water.

A wildcat well was drilled 1 mile east of town in 1981 in hope of finding oil. It penetrated 5,852 feet of Mesozoic and Paleozoic sedimentary formations before reaching the Precambrian basement rock. No oil.

Many buttes with flat tops rise 100 to 200 feet above the prairie south and west of Timber Lake. Resistant sandstone ledges within the Fox Hills formation cap their flat tops. Fossil shells of oysters and other marine organisms that lived here about 70 million years ago abound in some of the sandstone ledges.

Firesteel lies near the eastern edge of a belt about 40 miles wide in which the late Cretaceous Hell Creek formation is at the surface. High-quality black lignite coal lies about 70 feet above the base of the formation in the Isabel-Firesteel lignite field. Much of the upper Hell Creek formation has been eroded along the eastern edge of the outcrop area, so the coal there is at or near the surface. Early ranchers and homesteaders opened numerous wagon mines in the coal seam. With the coming of the railroad in 1910, production increased, reaching a peak of 50,000 tons per year during the early years of World War II. It dropped off sharply after the war as propane and other petroleum products became readily available.

Meadow was one of the first settlements in this part of the state. It lies near the eastern edge of a band of Tongue River clays that extends west to beyond the Slim Buttes. The clays were deposited during Paleocene time, perhaps about 60 million years ago. From high points along the highway on either side of Meadow, you can see all the way to White Butte, nearly 25 miles north.

Between Meadow and the Montana Line at Camp Crook

Just east of Bison are three large, shallow wind blowout depressions that contain water only during wet years. The town of Bison secures its municipal water supply from a series of wells that pro-

Meadow to Montana line (92 miles).

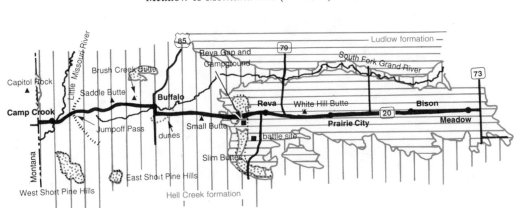

duce from aquifers in the Fox Hills formation at depths of 700 to 900 feet.

Clays belonging to the Chadron formation of the White River group cap White Hill Butte, 5 miles west of Prairie City and a mile north of the highway. A series of thin limestones in the upper 7 feet of the formation account for its resistance to erosion.

The Battle of the Little Bighorn was fought in southeastern Montana on June 25, 1876. General George Crook's troops arrived from southern Wyoming too late to participate in that action. He spent the rest of the summer rounding up scattered bands of Indians and encouraging them to return to their respective reservations. By early September, Crook found himself on the headwaters of the Heart River in western North Dakota, with his men and animals exhausted from fighting almost continuous rain, mud, and hunger. He detailed Captain Anson Mills to take a small party on a dash for the Black Hills to get supplies and fresh animals. They were to return to help the entire group to get to the northern Black Hills, where their next assignment was to protect the new gold-mining communities from Indians. On the way, Mills found a camp of 25 to 30 lodges of Sioux and Cheyenne Indians under the command of American Horse. In a surprise attack on the morning of September 11, he disrupted the

Petroglyph on sandstone at Slim Buttes. —S.D. Department of Tourism

village and destroyed the lodges and supplies. This skirmish was dig-
nified with the title "Battle of Slim Buttes." The actual battle site is
next to SD 79, about 8 miles south of the highway marker on SD 20.

The Slim Buttes form a long and narrow tableland that stands as
much as 600 feet above the prairie. They offer mute evidence of the
tremendous volume of material eroded from the Great Plains since
the caprock was laid down on a vast desert plain some 10 to 15
million years ago. The buttes are shaped like a boomerang in map
outline, with the long limb extending about 15 miles both north
and south from Reva Gap. At the southern end, the ridge bends
sharply to the southeast for an additional 10 miles. Two passes cross
the northern limb from east to west: Reva Gap, through which SD
20 passes, and JR Gap 8 miles farther south. SD 79 crosses the southeastern limb.

Reva Gap exposes the rocks of the Slim Buttes. The forest service
campground makes a convenient base for exploring the Castles and
Battleship Rock area on foot. Away from the base of the buttes, the
bedrock is Ludlow formation of Paleocene age. The Tongue River
beds that must once have covered it were eroded before the pale
clays and silts of the White River formation were laid down. They
are what you see exposed in the buttes.

Detail of Tertiary rocks at Reva Gap.

Age	Formation		Description
Miocene	Arikaree		Fine white sandstone with much volcanic ash; 260 feet
Oligocene	White River group	Brule	Tan-to-pink clay and silt, with zones of calcareous nodules and channel sandstones; 350 feet
		Chadron	Typical greenish clays like those in Big Badlands; 7–15 feet Dazzling white beds, coarse white sand and conglomerate; 10–40 feet Golden brown beds, iron-stained clays; 40–70 feet
Paleocene	Fort Union	Tongue River	Brown-to-tan clay, locally missing by erosion
		Ludlow	Dark gray, carbonaceous clays, sandstones, and lignites; bottom not exposed

Erosion of White River Oligocene beds at Slim Buttes. —S.D. Department of Tourism

Those weak Ludlow clays failed in a band of landslides around the foot of the buttes. Watch for evidence of this slumping in the tilted rocks and undrained depressions at the campground. Pits of several early coal mines in the Ludlow formation are within less than a mile of Reva Gap.

The White River beds at the northern end of the Slim Buttes are broken along faults and tilted to steep angles. Similar faults exist around other groups of high buttes in Harding County. Many geologists have tried to explain this, but none have succeeded. In 1895, one geologist argued that the sharp folding was followed by deep erosion between Oligocene and Miocene time. The next suggestion, made in 1911, was that the inclined beds are cross-bedding of the type that develops on the advancing face of a sand dune or the front edge of a shifting sandbar. That leaves unanswered the question of why such large cross-bedding should occur only in such a very restricted area.

A third suggestion is that the folding is part of a regional event. The absence of disturbed beds at depth, as seen in seismic studies made during a search for oil, cast doubt on this hypothesis. A fourth

An angular unconformity. Brule beds were tilted, eroded off, and then covered by younger Arikaree formation. —Robert Wilson

suggestion would explain the tilted blocks as slumps at the face of a steep cliff being undercut by a stream between Oligocene and Miocene time. However, the older rocks show no evidence that such a deep stream channel ever existed. A recent suggestion attributes the faulting and tilting to some type of failure in the very weak clays of the Ludlow formation, permitting the White River beds to drop and tilt. The field for speculation and interpretation is still wide open. Poke around and formulate your own theory.

The first written record of the Slim Buttes is in the journals of the Astorian expedition, led by William Pierce Hunt in 1811. The next is in the journal of the Ashley fur-trading expedition of the upper Missouri country in 1823. In 1978, an area of 1,005 acres, including the Castles and Battleship Rock, was registered as a National Landmark.

For about 5 miles east of the intersection with US 85, the road crosses a belt of fine, windblown sand. Small dunes and wind blowouts stabilized by prairie grass make a hummocky surface.

Buffalo, at the junction with US 85, is a trade center for northwestern South Dakota, southeastern Montana, and southwestern North Dakota. Buffalo relies for its municipal water supply on wells that produce from sandstone aquifers within the Hell Creek formation at depths of 120 to 130 feet.

Small butte of Hell Creek clay west of Slim Buttes.

Dissected Hell Creek beds south of SD 20 in the Jumpoff Country. —Robert Wilson

Except for the tips of some of the higher buttes north of the highway, bedrock between Buffalo and Camp Crook is clay of the Hell Creek formation. Saddle Butte, north of the highway 9 miles west of Buffalo, offers an exposure of about half of the formation thickness. Outcrops on the east side of Jumpoff Pass expose still higher Hell Creek beds. The pass marks the drainage divide between short streams that flow west into the Little Missouri River and the broad headwaters of the Grand River to the east. The elevation at the pass is about 3,370 feet.

Jumpoff Pass provides a good place to examine a badlands wall, an escarpment marking the eroding edge of an upland area. It is between 100 and 200 feet high and is very steep in most places. The gentle slope of the east-facing escarpment at the pass makes a road crossing feasible. The edge of the upland surface retreats to the south and west as erosion eats into it. Many small buttes left behind make a belt of very rugged country parallel to the wall. The escarpment and the rough country next to it are locally called the Jumpoff Country.

The West Short Pine Hills, 7 to 15 miles south of Camp Crook, are a ridge of buttes that support a nice cover of trees. They stand several hundred feet above the prairie and resemble the East Short Pine Hills described in the log for US 85.

West Short Pines Gas Field

Although oil prospecting in the deep Paleozoic rocks started in the 1930s, no one paid any attention to the shaly upper Cretaceous rocks. In 1978, gas was discovered in the thin Eagle sandstone within the Pierre shale at depths of 1,500 to 2,000 feet. The sandstone is correlated with the productive Shannon sandstone of Wyoming. Drilling to date has outlined a field 8 miles long, from north to south, and 3 miles across. Well spacing is one for each 320 acres. Yield of individual wells is not large, but proximity to a pipeline extending from North Dakota to Wyoming makes development economically feasible. This is the only sizable gas field in South Dakota.

Little Missouri River

The Little Missouri River is a small stream in a big valley. You can spot it by the line of cottonwood trees. The river flows from northeastern Wyoming across the southeastern corner of Montana and the northwestern edge of South Dakota and finally enters the Missouri River in northwestern North Dakota. Before late glacial time, the Little Missouri River drained a large area in northeastern Wyoming. When the Cheyenne River began to flow in western South Dakota, it had a high gradient and eroded rapidly headward. The north fork of the Cheyenne, now called the Belle Fourche River,

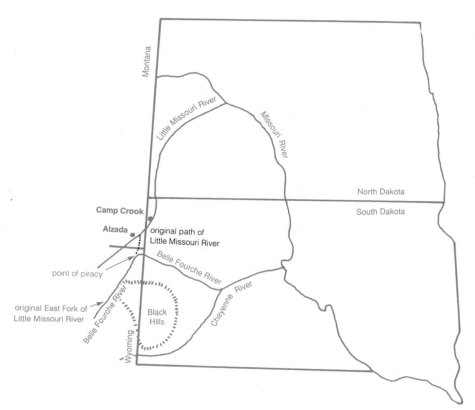

Piracy of the upper part of the Little Missouri River by the Belle Fourche River.

rapidly extended its headwaters northwestward. It intercepted and captured the principal flow of the Little Missouri River near Alzada, Montana.

Camp Crook

Camp Crook is on the west bank of the Little Missouri River; it was founded in 1883 under the name of Wickhamville. During the winter of 1882–83, General Crook placed a detachment of troops at the settlement to guard a work crew cutting railroad ties for the Northern Pacific Railroad. In 1885, the name changed to Camp Crook. The town finally incorporated in 1908.

A stretch of gravel road extends 4 miles west from Camp Crook to the Wyoming line. One small butte of Hell Creek clay breaks the monotony. At the state line, a gravel road leads north 10 miles to a

Little Missouri River at Camp Crook.

large erosional remnant of White River beds called Capitol Rock. From the proper vantage point, it does indeed resemble a classic domed capitol.

SD 34
Fort Pierre–Sturgis
175 miles

All the bedrock along this highway is soft sedimentary formations deposited in the shallow inland sea during late Cretaceous time. Most of it is the dark gray Pierre shale, but remnants of the overlying Fox Hills formation survive in a few places. The Fox Hills formation is mostly muds and sands deposited in coastal lagoons as the shoreline made its final retreat across the state.

Between Fort Pierre and Plainview

The highway between Fort Pierre and the junction with SD 63 follows the rolling upland surface of the divide between the Cheyenne River on the north and the Bad River on the south. Bedrock is all dark gray Pierre shale. You see more Pierre shale between SD 63 and Billsburg.

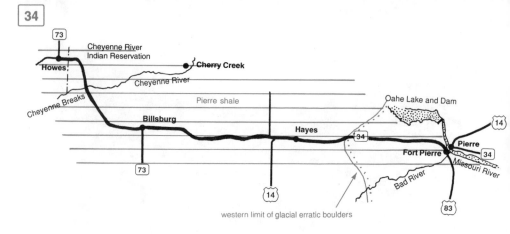

Fort Pierre to Howes (96 miles).

The Diamond Ring Ranch is 12 miles west of Hayes and south of the highway. In 1959, the owners studied the geology of all the nearby oil tests, then decided to drill a deep artesian well to the Madison formation on a high divide between three drainages. The idea was to obtain a flow of hot water that could be diverted into any of three large pastures. It would flow for some distance before freezing, so the cattle could have warm water to drink even in the coldest weather. It worked. The well, which is 4,112 feet deep, flowed more than 250 gallons of water per minute at a temperature of 152 degrees Fahrenheit. Total dissolved solids were slightly more than 2,000 parts per million. Fifteen years later, they equipped the well to extract heat from the water for drying grain and heating several ranch buildings.

Bedrock between Billsburg and Plainview is Pierre shale, deeply dissected by the Cheyenne River. The weak shale makes unstable slopes that slump in many small landslides. Watch for the hummocky topography. The rugged breaks support a sparse growth of junipers, but very few deciduous trees or shrubs.

Between Plainview and Sturgis

Between Plainview and Union Center, the highway crosses a remnant of Fox Hills formation overlying the Pierre shale; watch for the signs in the landscape. The Fox Hills formation weathers to sandy soil much better than that developed on the Pierre shale. More deciduous trees appear in the draws, and the shelterbelt tree plantings have survived better than those planted on the Pierre shale. Shallow sandstones within the Fox Hills formation offer groundwater far superior to that found in the Pierre formation.

Stoneville is about 15 miles north of Union Center. A homesteader digging a well in the Fox Hills formation in the early 1920s discov-

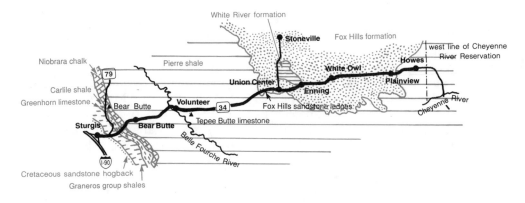

Howes to Sturgis (80 miles).

ered a bed of lignite coal suitable for domestic use. Test holes showed that the coal lay 30 to 55 feet below the surface, depending upon the topography, and that the seam ranged from 30 to 52 inches thick. Several small mines, mostly northeast of Stoneville, supplied coal for local use. By 1936, only one mine continued to operate, with an annual output of 600 tons. Mining ended with World War II.

Coal is rare in the Fox Hills formation. Apparently the shoreline of the Cretaceous inland sea remained in one place long enough in the Stoneville area for coal swamps to develop along the coastal flats and estuaries.

From 2 miles west of Union Center to a point due south of Bear Butte, the highway is on Pierre shale. The breaks of the Belle Fourche River, originally known as the North Fork of the Cheyenne River,

Spherical concretions in Fox Hills sandstone form as calcium carbonate is deposited around a central nucleus. These are south of the highway near Enning.

expose 200 feet or more of Pierre shale. Cutting of the deep Missouri River trench through central South Dakota during the ice age also caused the tributary streams to erode their channels. Here, at Volunteer Crossing, the deepening was about 200 to 250 feet. The Belle Fourche River not only deepened its valley, but also eroded headward and eventually captured the headwaters of the Little Missouri River in southeastern Montana. This added the drainage of a large area in northeastern Wyoming and southeastern Montana to the flow of the Belle Fourche River, further accelerating its rate of erosion.

Tepee Buttes Limestone

Just below the crest of the east side of the Belle Fourche River Valley, a mass of tepee buttes limestone lies a few tens of feet south of the highway. It is about a quarter mile west of the entrance to the Leer Ranch, or 3.2 miles east of the Belle Fourche River bridge. The mass, 5 or 6 feet across, contains shells of the clam *Lucina occidentalis*, which lived in seawater. As erosion continues, this limestone may eventually form the peak of a conical butte. See it now while you do not have to climb a butte. Some geologists believe that these patches of limestone formed around warm springs on the seafloor.

Bear Butte

Bear Butte rises more than 1,200 feet above the adjacent prairie. It was an early landmark and a sacred place for the Cheyenne Indians. The higher portion is now a state park. A relatively easy foot trail leads to the top from a visitor center on the south flank.

The butte is an igneous intrusion that rose as molten magma through the gently dipping sedimentary strata. The main intrusion is a very fine-grained rock with a chemical composition similar to granite. It has a porphyritic texture, which means that some of the mineral grains are much larger than others. An age date reveals that it is 51 million years old, about the same age as many other igneous intrusions in western South Dakota and nearby parts of Wyoming. The high slope on the eastern side is a block of Mississippian Madison limestone, tilted upward 4,500 to 5,000 feet by the intrusion.

From a point due south of the butte to the water gap at Sturgis, the highway crosses successively older rocks tilted up along the northeastern flank of the Black Hills. Formations that you cross, mostly concealed by alluvium and terrace gravels, include the Niobrara chalk, the Carlile shale, the Greenhorn limestone, and the Belle Fourche and Skull Creek shales. These dark gray sediments total more than 1,600 feet thick. All were deposited in shallow seawater.

More conspicuous are the erosional terraces along the south side of Bear Butte Creek. The highway alternates between the present

Bear Butte from the south.

floodplain and the next higher or Sturgis level. The still-higher Rapid terrace appears on the skyline both north and south of the creek.

Fort Meade

Fort Meade was founded in 1877 as the home of the Seventh Cavalry after the disastrous Battle of the Little Bighorn the previous June. It was an active cavalry post until World War II, when the cavalry moved to Louisiana and mechanized. They never returned to Sturgis. The Veterans Administration now uses the old facilities, greatly expanded and modernized.

When Fort Meade was established, a large surrounding area of hills and prairie was set aside to provide timber, building stone, water, pasture, and hay. The more accessible part is now a recreational area. A good gravel road turns south at the main entrance to the post, with an arrow pointing to the Fort Meade National Cemetery. The road wanders through the foothills, cuts through the next water gap to the south, and joins I-90 at the Black Hills National Cemetery. The total distance is only 4 miles, and it passes along sandstone outcrops that support pine trees and stream valleys that contain a rich assortment of deciduous trees and shrubs. It is a shortcut to Rapid City.

Cretaceous Sandstone Hogback

The highway enters Sturgis through the narrow valley that Bear Butte Creek cut through the Cretaceous sandstone rim of the Black

Hills, which is about 600 feet high. The following geologic section is well exposed:

FORMATION	THICKNESS
Fall River sandstone: massive at base, slabby upward	35 feet
Lakota formation	
Fuson member: variegated shales, clays, and sands	100 feet
Chilson member: massive sandstone	235 feet

Beneath these are about 400 feet of Jurassic shales and sandstone, and beneath them are about 700 feet of poorly exposed red sandstone and mudstone deposited during Permian and Triassic time. To get a close view of these important sandstone aquifers, turn north on Junction Street in downtown Sturgis and drive less than a mile along their outcrop to the top of Sly Hill. A turnaround at the top, at the Fort Meade Reservoir, offers a fine view of Bear Butte and of the old military post.

Lakota sandstone cliffs on Sly Hill, Sturgis.

SD 44
Interior–Rapid City
75 miles

The eastern half of this route crosses pale sediments of the White River group. They were deposited on an arid alluvial plain during Eocene and Oligocene. The western end crosses dark gray marine shales partly buried beneath recent stream sediments in the valley floors, and beneath terrace gravels in the higher areas.

Between Interior and Scenic

Between Interior and the ghost town of Conata, the highway crosses a low and much eroded area next to the White River. In the lowest spots, erosion has removed the pale sediments of the White River group, exposing the Pierre shale and locally exhuming an ancient landscape that was eroded on the Pierre shale during Eocene time.

Buried Eocene Soil, a Souvenir of a Tropical Time

Paleontologists searching for vertebrate skeletons a century or more ago noticed an interval of red and yellow clays overlying the gray Pierre shale but below the light greenish gray beds of the lower bad-

Interior to Rapid City (75 miles).

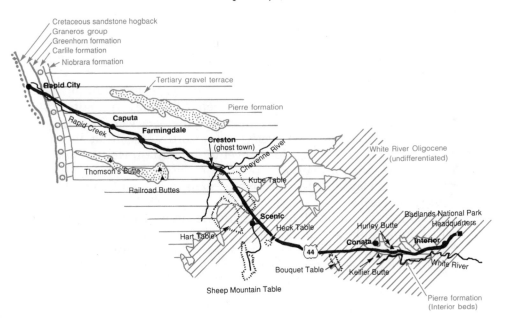

lands clays. In 1923, Freeman Ward, the first state geologist, named this colored zone the Interior beds because of their good exposures near that town. Geologists later recognized that these were in fact the Pierre shale weathered under warm and wet conditions to a red tropical soil.

As is so obvious, modern weathering of the Pierre shale produces dark gray gumbo clay about the same color as the parent rock. Things went differently under the radically different climate of Eocene time: Weathering oxidized the iron pyrite in the shale, creating shades of red and yellow for several tens of feet below the land surface. This colorful fossil soil survives where the clays of the Chadron formation buried it during Oligocene time. Recent geologic mapping in western South Dakota shows that similar red and yellow soils developed on several of the older shales exposed at the same time.

West of Interior, but not visible from the highway, a conspicuous bed of white sandstone separates the buried red and yellow soil from the overlying green clays of the Chadron formation, part of the White River group. Although the white sandstone contains no animal fossils, it does contain fragments of petrified wood that must have floated in as scraps of wood, became buried in the sand, and then petrified. The sandstone is locally a good aquifer, but its distribution is unpredictable.

The White River Group

Between Conata and Scenic, you see a full section of late Eocene and Oligocene clays and silts that belong to the White River group of formations. Streams and wind deposited the sediments on the early Eocene erosion surface, which had local relief of at least 150 feet. So the thickness of the Chadron beds is highly variable, ranging from just a few feet over the old knobs to more than 150 feet in the old low areas.

The Chadron formation typically weathers to low hummocks, locally called haystacks, but never to steep cliffs. The formation contains bentonite clays that swell up when they get wet. The upper few inches of the outcrop develops a popcorn surface that absorbs water nearly as fast as the clouds ever dump rain on western South Dakota. The popcorn surface appears dry just a few hours after a rain, but the clays just beneath may remain saturated and slick for days, even weeks.

The most famous fossil in the Chadron beds is of a titanotherium, which looked like an early prototype of a rhinoceros. It stood 8 feet high at the shoulders and was as much as 14 feet long. Its teeth were 2 to 3 inches across. You are likely to see glistening fragments

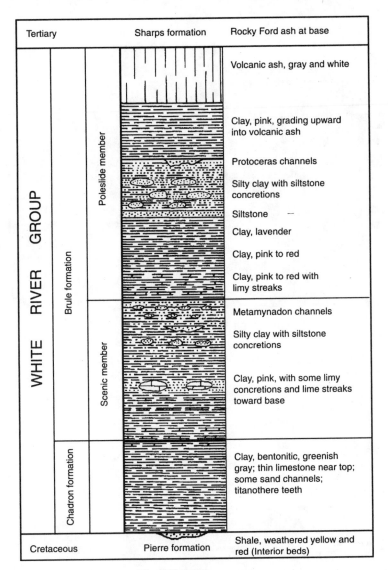

Tertiary			Sharps formation	Rocky Ford ash at base

The White River group.

of tooth enamel scattered across many outcrops of the Chadron formation.

Two or three thin and discontinuous zones of resistant limestone mark the break in slope between the low, rolling topography of the Chadron formation and the steep cliffs of the overlying Brule formation. Algae deposited much of the limestone, and it contains fossil clams and snails that lived in shallow lakes.

Titanotherium skeleton. —S.D. Museum of Geology

Cliffs of Brule formation loom over rounded haystacks of softer Chadron beds.
—R. W. Wilson

The Brule formation, the upper part of the White River group, was named for the Brule band of Sioux Indians, not for the customary geographic locality. It is several hundred feet thick and weathers to steep bluffs. Good exposures show ledges of sandy gravel between banded silts and clays. These mark channels of streams that wandered over the old alluvial plain as it was accumulating sediment. Geologists have divided the Brule formation into several smaller units based on color, rock type, and presence of nodular concretions. These units trace continuously for at least 50 miles. The Brule formation thins eastward, away from the western source of the sediments. The flanks of some of the higher tables in the area around the town of Scenic expose about 450 feet.

When it rains, parts of the Brule formation shed water like a roof, which explains the countless little stream channels that typically dissect the badlands outcrops. Its ability to shed surface runoff also explains why badlands exposures of the Brule formation look so different from those of the Chadron formation.

Occasionally you may see vertical dikes several inches thick cutting through the Brule formation. In some places, you can follow them for several tens of yards. Unlike most dikes, these are not igne-

Cliffs and columns of Brule clay protected from erosion by cemented channel sandstones. —N. H. Darton, U.S. Geological Survey

Chalcedony veins or dikes weathering out of popcorn-weathered badlands clay. —S.D. Museum of Geology

ous rocks; they are sandy clay. Apparently, the soft clay flowed into cracks that opened as the formation dried out after it was deposited. The dikes are generally harder than the surrounding clays and commonly support the crests of sharp ridges.

Vertical veins of dark gray chalcedony, a translucent variety of extremely fine-grained quartz, also exist. They are thin and generally hard to see. Their broken pieces make the blades of dark chalcedony that are scattered across the Brule formation in several areas between Interior and Scenic. These veins formed as fillings in vertical shrinkage cracks within the Brule clays. The thin plates are virtually indestructible and remain behind as a protective shingle as the surrounding clays erode. Prehistoric Indians chipped a sharp edge on one side, then used them for knives and scrapers. People find chalcedony artifacts several hundred miles in all directions from this rather limited source area.

The Big Badlands

Between 2 and 5 miles east of Scenic, the highway crosses a narrow strip of the Badlands National Park. It connects the main tourist area southeast of Wall with Sheep Mountain Table and a large area farther south that the federal government made into a bombing range during World War II.

Scenic is in a large depression known as the Scenic basin. The town was a small but important ranch supply and cattle shipping center established before the Chicago, Milwaukee, and St. Paul Rail-

Toadstools, concretions weathering out of the Brule formation, stand on columns of clay.

road built through to Rapid City in 1907. The town gets its water from a large shallow well in gravel at the south end of Kube Table 2 miles north of town. The well is 100 feet higher than Scenic, so gravity delivers the water.

You can see spectacular views of the Big Badlands from the top of Sheep Mountain Table. The Big Badlands refers to the spectacular erosion features lying between the Cheyenne and White Rivers in western South Dakota. Drive south from Scenic 4 miles on Pennington County Road 589, turn west for 3 miles on an unpaved road, then ascend 350 feet in less than a half mile to reach the top of Sheep Mountain Table. The Big Badlands are an erosional remnant of an old land surface that survives between the Cheyenne and White Rivers. The high southern end stands nearly 400 feet above the surrounding lowland. There, a thick layer of white Rocky Ford volcanic ash caps the Brule formation. Weathering of the ash created vertically fluted structures quite unlike the horizontal banding in the underlying clays.

Between Scenic and Rapid City

The highway leaves the Scenic basin to climb onto Kube Table, another remnant of the old erosion surface that the Cheyenne River and its tributaries carved. A sheet of gravel, presumably laid down in Pliocene time, covers the table, and a thin blanket of windblown dust covers that. Bedrock beneath all of that is of the White River

Erosion into pinnacles at south end of Sheep Mountain Table. White layers are nearly pure ash.

group at the south end, with Pierre shale at the north end. At the south end of Kube Table, a short drive leads to a good view of the Scenic basin and the Big Badlands to the south.

At the northwestern end of Kube Table, the road drops rapidly through the Pierre shale to the Cheyenne River bottom. The roadside ditches are a good place to collect fossil crabs from the Pierre shale. Look for tan limestone nodules, about the size of your thumb, with embedded fragments of shiny black crab shell.

Since it began to deepen its channel sometime during the ice ages, the Cheyenne River has eroded a valley 250 feet deep and more than a mile wide. The steep valley slopes exhibit countless slumps and small landslides on the weak and unstable Pierre shale. Gravel pits within the floodplain furnish much of the sand and gravel used in construction work in the Black Hills area.

Early records show that a fur-trading post was established at the junction of Rapid Creek and the Cheyenne River in the winter of 1829–1830. The enterprise ended abruptly in late January 1832, when someone dropped a lighted candle into a keg of black powder. Later floods destroyed any evidence of the old site, and recent efforts to locate it have failed. The South Dakota State Historical Society placed a marker to note the Oglala Fur Post.

The highway mostly follows the north side of the valley of Rapid Creek, following ancient stream terraces wherever possible. This route offers a chance to see the terraces that dominate the landscape around the periphery of the Black Hills.

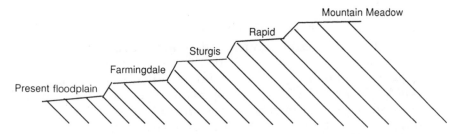

Profiles drawn at right angle to streams. Succession of gravel-capped terraces abandoned by Black Hills streams as they alternately deepened and widened their valleys. Downcutting started soon after the close of the Oligocene epoch, but the lower terraces could not have been formed until the Missouri River trench was developed in mid-glacial time.

Cross-valley profiles of streams leaving the Black Hills show quite a series of terraces. Some geologists contend that when the Black Hills were rising, streams deepened their valleys; when they were stationary, streams widened their valleys. Terraces are obvious along all of the streams. Rapid Creek has slowly shifted its course to the south, so the terraces are mainly on the north side of the valley.

The Farmingdale terrace is the first above the floodplain of Rapid Creek. The highway uses its uniform grade much of the way between Creston and Farmingdale. West of Farmingdale, the highway shifts briefly onto the next higher Sturgis terrace, then drops to the present floodplain for most of the way to Rapid City. The highway hugs the base of the slope to save good farmland.

The runways of the Rapid City Municipal Airport are on the Rapid terrace, the second erosion level above the Farmingdale terrace. It lies about 160 feet above the present floodplain.

Thomson's Butte

Thomson's Butte is 5 miles south and a bit west of Caputa. It is an erosional remnant of the White River formation perched on the divide between Rapid and Spring Creeks. More than 280 feet of Chadron and Brule clays rest on Pierre shale, and they are in turn capped by 20 feet of coarse gravel. The butte was named for an early rancher, Thomas Thomson. His son, Albert, began as a manual laborer working for the American Museum of Natural History, and finally became a professional fossil collector and preparator. He went to the Gobi Desert with Roy Chapman Andrews in the 1920s, where he collected the first recorded nest of fossil dinosaur eggs.

Niobrara Chalk

The auto racetrack east of Rapid City is on the chalky shales of the Niobrara formation. Thin and discontinuous layers of hard limestone within the chalk contain countless shells of small oysters. Nearly spherical limestone concretions as much as a foot across also exist. In the few years since the vertical face was cut, the outcrop has weathered from bluish gray to the more familiar yellowish orange. It exposes the lower half of the Niobrara formation, which is chalkier and less shaly than the upper half. The total thickness of the Niobrara formation in western South Dakota is about 200 feet.

Rapid City

The highway follows the floodplain of Rapid Creek through town. Remnants of two old erosion levels are conspicuously evident in Rapid City. The higher is the Rapid terrace, a narrow upland surface in the southeastern part of town that early settlers called Signal Heights. It stands about 200 feet above the modern floodplain. Look for it in the ridge with the flat top behind the School of Mines and in the Star Village development east of Fifth Street. In the northeastern part of town, part of East North Street is on the Sturgis terrace, as is the National Guard Camp in West Rapid City. It stands roughly 70 feet above the modern floodplain of Rapid Creek.

Rapid Creek has been furnishing irrigation water since the first settlers arrived. Formal water rights claims date back to 1877. Leakage from ditches and excessive use of water have raised the water table in the valley bottom to within a foot or two of the surface.

SD 63
US 18–North Dakota Line
232 miles

The route crosses outcrops of sedimentary rocks, some deposited in seawater, others on land. They range in age from late Cretaceous to Oligocene and represent a time span of about 50 million years. The road crosses gently rolling to flat uplands, deeply dissected at intervals by geologically very young breaks along the major streams, which flow east. Erosion of the soft rocks has created a rather monotonous landscape.

Between US 18 and Midland

Between US 18 and the White River breaks, a distance of about 40 miles, the route crosses pale clays and sands of the White River group. They were deposited during late Eocene and Oligocene time. The

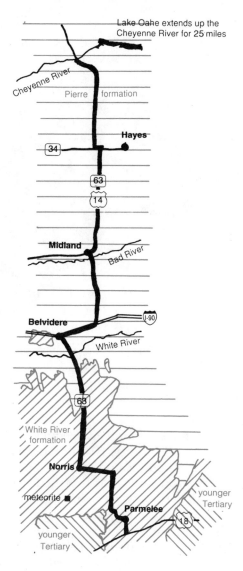

*US 18 to Cheyenne
River (117 miles).*

beds are not as fossiliferous as the thicker section to the west in Badlands National Park.

North of Norris, the highway follows the drainage of Black Pipe Creek to the White River. Ten miles north of Norris, bedrock changes from the pale clays of the White River group to the dark gray shales of the underlying Pierre formation.

Between White River and Midland on the Bad River, bedrock is all Pierre shale except for a small area at Belvidere, where a remnant of pale badlands clays caps the drainage divide. Water from the deep Midland wells is warm enough to heat two school buildings before it goes into the city mains.

This meteorite, found in Bennett County, measures more than 15 inches long. The fused surface and pits formed during heating on passage through the atmosphere. —S.D. Museum of Geology

Meteorite

In 1934 someone found a large nickel-iron meteorite on the Dale Ranch in Bennett County, 10 miles south and 3 miles west of Norris. It weighed 195 pounds and consisted of a coarsely crystalline variety of nickel-iron known as hexahedrite. The specimen, with one surface polished to show its geometric internal structure, is on display in the Museum of Geology at the School of Mines in Rapid City.

Polished and etched surfaces of nickel-iron meteorites typically show a spectacular pattern of large crystals of metal arranged in intricate arrays of triangles. That pattern is diagnostic; it resembles nothing you might find in an ordinary piece of scrap iron. Standard procedure in identifying nickel-iron meteorites is to saw off a slice, polish the sawed surface to a mirror finish, then lightly etch it with diluted nitric acid. The acid makes the crystal pattern very easy to see.

Between Midland and the North Dakota Line

North of Midland for about 40 miles, SD 63 crosses a gently rolling erosion surface between the Bad and Cheyenne Rivers. The highway crosses the Cheyenne River at the upper end of an arm of Lake Oahe, which extends roughly 25 miles upstream from the Missouri River.

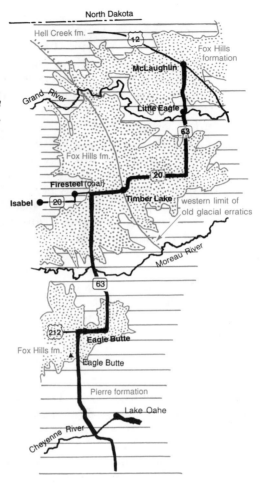

Cheyenne River to North Dakota line (117 miles).

The hummocky surfaces of the steep valley slopes of each of these major streams illustrate the way the Pierre shale weathers and erodes. The shale is unstable, particularly when saturated, and slumps in countless small landslides that carry the soil to the valley bottoms, where streams haul it away. In the short term, one of the Missouri River dams will trap the sediment, but its ultimate destination is the Mississippi River Delta in the Gulf of Mexico. North of the Cheyenne River, the Cheyenne River Indian Reservation extends 33 miles north to the southern border of the Standing Rock Reservation.

Between Eagle Butte and the state line, sands and clays of the Fox Hills formation cap the uplands; the streams have eroded their valleys deeply into the Pierre shale beneath that cap.

The town of Eagle Butte is on the broad divide between the Cheyenne and Moreau Rivers. A remnant of Fox Hills formation about 6 miles wide caps the divide at the town site. Eagle Butte is the administrative headquarters for the Cheyenne River Indian Reservation. Originally, the administrative headquarters was at Cheyenne Agency along the Missouri River—a heritage from the days when steamboats plying the upper Missouri supplied all military posts and Indian agencies. When the Oahe Dam was built in 1956, the agency moved to the higher, more central location at Eagle Butte. A well was drilled to obtain water from the Madison limestone at a depth of 4,325 feet. The well flowed about 110 gallons per minute at a temperature of 127 degrees Fahrenheit. But the water tasted bad, so a pipeline was built a few years later to bring water from the Missouri River.

It is no accident that the towns along this part of the route are on the Fox Hills upland. Railroads, long abandoned, chose the flat drainage divides for their branch lines to the west. Early settlers soon realized that the range grass is better on the outcrops of Fox Hills sandstone than on the formations above or below it. And good groundwater is generally available there at a depth of less than 150 feet.

Lenses of cemented sandstone within the Fox Hills formation cap low buttes to the south between Timber Lake and Glencross. These sandstones commonly contain abundant shells of oysters and other animals that lived in the shallow inland sea during Cretaceous time. In 1981 a large oil company drilled two deep wildcat wells between Timber Lake and Glencross. Both penetrated the ancient granite basement more than a mile below the surface. Neither reported shows of oil or gas.

Weathered granite boulders scattered about east and north of Timber Lake show that one of the earlier glaciers once reached at least 35 miles west of the present Missouri River. Erosion has since removed all glacial debris except an occasional boulder.

Sandstone-capped buttes southwest of Timber Lake.

Little Eagle is in the Standing Rock Indian Reservation on the floodplain of the Grand River. The Indians generally settled where the river breaks protected them from the wind and where wood and surface water were readily available. Normally dry sloughs on the floodplain west of town mark abandoned meanders of the Grand River. A few miles upstream from Little Eagle was the camp of Sitting Bull. He and several of his followers were killed while being placed under arrest during the messiah craze in 1890.

Settlers founded McLaughlin shortly after the Chicago, Milwaukee, and St. Paul Railroad bridged the Missouri River at Mobridge in 1906. It was named for Colonel James McLaughlin, who was Indian agent at Standing Rock while Sitting Bull lived on the reservation. Two wells drilled 160 feet into the Fox Hills sandstone furnish the town with water. For a few miles south of the state line, the road crosses the somber gray shales of the late Cretaceous Hell Creek formation.

SD 73
Nebraska Line–North Dakota Line
243 miles

On this traverse across the full width of South Dakota, we see only sedimentary rocks, still lying as flat as when they were brand new. The older rocks were deposited during Cretaceous time as muds and sands were laid down on the floor of the shallow inland sea; they come in somber shades of medium to dark gray. The younger rocks were deposited during Tertiary time on a broad alluvial plain; they are generally pale gray or yellowish gray. Most of the rocks are soft and easily weathered, so they erode into gently rolling plains.

Between the Nebraska Line and Philip

The route between the Nebraska line and the White River crosses the Pine Ridge Indian Reservation, the largest of seven for the tribes of the Sioux Nation. It is the home of about 8,500 Oglala Sioux.

From the state line to the upper reaches of the White River breaks, bedrock consists of soft and poorly cemented pale sandstone and clay, which weather to a gently rolling landscape nearly without distinctive landmarks.

The Sandhills, which cover much of northwestern Nebraska, here extend about 7 miles into South Dakota. They are windblown sand, originally moving dunes, but now are largely stabilized under a cover

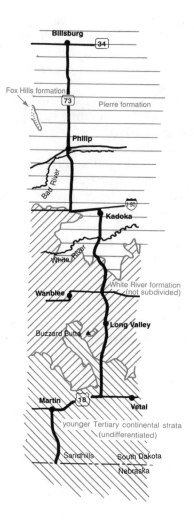

Nebraska line to Billsburg (120 miles).

of grass and brush. The sand probably weathered from weakly cemented outcrops of the Valentine formation farther west.

Immediately north of the Sandhills, the road drops into the valley of Lake Creek, where you can see outcrops of the sandy Valentine formation. It was deposited during late Oligocene time, after a period of erosion. The Valentine formation fills valleys in the erosion surface. That explains why it is highly variable in elevation, thickness, and distribution. North of the sand dune area, the road cuts successively down through Miocene and Oligocene sedimen-

180

Sandhills, south of Martin.

tary formations and into the underlying dark Cretaceous shales in the breaks of the White River.

Martin is on an upland surface underlain by the Miocene Harrison formation, a fine-grained sandstone full of concretions and cemented ledges. Martin gets its municipal water from several wells drilled some 500 feet in Tertiary strata. The water appears to come from fractures in the rock, rather than from storage space between the sand grains within specific layers.

A few miles north of Martin, the landscape changes from gently rolling prairie in the area of Miocene outcrops to spectacular badlands topography developed in the White River formation along the northern side of the Indian reservation.

Although the White River here flows in a valley cut in dark gray to black Pierre shale, it starts in a broad area of pale clays. Erosion in the headwaters is very active, so the main stream carries a large load

Sedimentary rocks in the vicinity of Martin.

Recent		Sandhills		
Miocene	Valentine fm.		Sand; fine and gray	
	Harrison fm.		Sand; fine, gray to olive, with concretions and cemented ledges	150 feet
Oligocene	Monroe Creek fm.		Sandstone; fine, pink to brown	250 feet
	Sharps fm.		Siltstone; buff-to-pink Rocky Ford ash at base	150 feet
Eocene	White River group		Clays; greenish gray to tan with concretions, ledges, and channels	300 feet
Cretaceous	Pierre fm.		Dark gray marine shale	800 feet

of nearly white clay. That is why early travelers called it the White River. The town of White River gets its water from wells drilled less than 25 feet into the stream deposits in the floor of the valley.

Kadoka, on the divide between the White River and the Bad River, is on Pierre shale. Before 1950, the town got its municipal water from Kadoka Lake southwest of town. It now depends on artesian water from the Dakota sandstone, at a depth of 2,700 feet.

Philip is in the valley of the Bad River, here entrenched about 250 feet below the general prairie surface. The sides of the valley are hummocky because they are full of landslides. The slopes are unstable partly because they are on Pierre shale, partly because the valley walls are steep, one result of the rapid erosion during the ice ages.

Philip gets its water supply from the Madison limestone at a depth of 4,200 feet. Water from that depth comes to the surface at a temperature of 154 degrees Fahrenheit. The hot water from another Madison well heats the high school building. Water from shallower artesian wells in the Dakota sandstone heats several other buildings.

Between Philip and the North Dakota Line

The route north of Philip traverses an old erosion surface developed on the Pierre shale. Watch for the hummocky valley walls where

Placenticeras, related to the modern pearly nautilus. The outer shell is removed to display the intricate crenulations of the septa, which divide the shell into chambers. This is about 8 inches across. —S.D. Museum of Geology

the road crosses the deep valley of the Cheyenne River. That is typical landslide topography, the kind that generally develops on steep slopes on the Pierre shale.

A few miles south of Faith, the highway crosses an inconspicuous outcrop of the Fox Hills formation. This is an old beach, a deposit laid down during the retreat of the inland sea that flooded most of South Dakota during late Cretaceous time. It separates the Pierre shale from the alluvial plain deposits of the overlying Hell Creek formation upon which Faith is built.

Billsburg to Lemmon (124 miles).

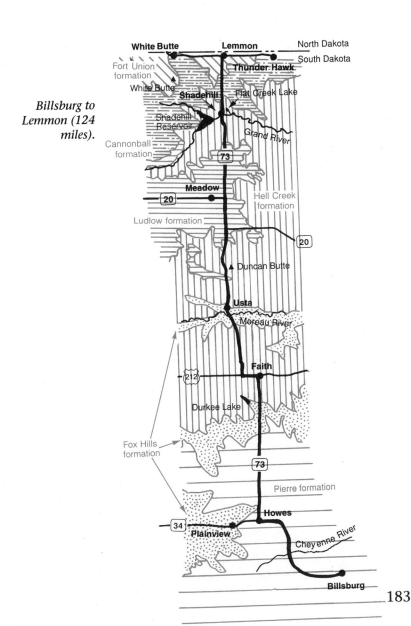

Faith, formerly the western terminus of a branch line of the Chicago, Milwaukee, St. Paul, and Pacific Railroad, takes its water supply from Lake Durkee, a reservoir a few miles south of town. Bedrock between Faith and Usta is the somber clays of the late Cretaceous Hell Creek formation.

The Moreau River is shown as the Owl River on some old maps. It is one of the major streams that drain northwestern South Dakota. Here, the valley of the Moreau River cuts through the Hell Creek formation and into the beach sands and clays of the Fox Hills formation. Usta is in a big meander of the Moreau River, just below the point where Rabbit Creek enters the main stream. Usta is a corruption of a Sioux nickname applied to the pioneer cattleman Ed Lemmon.

Between Usta and Shadehill, the bedrock is late Cretaceous Hell Creek clay in the bottoms, somewhat lighter gray sands and clays of the early Tertiary Ludlow formation on the uplands. You cannot tell them apart from the car window. Ludlow clays also cap Duncan Butte, which stands 145 feet above the prairie on the east side of the road.

The small settlement of Shadehill was founded in 1916. The steep grade into the river bottoms was known as Shade Hill, for the local highway engineer, M. L. Shade. Shadehill Reservoir is behind an earthen dam across the Grand River, which the Bureau of Reclamation built as a flood control project in 1951. The dam is just below the junction of the north and south forks of the river, so the reservoir has a map outline that suggests a wishbone. Bedrock at reservoir level is the impervious Hell Creek formation; the overlying beds

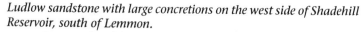

Ludlow sandstone with large concretions on the west side of Shadehill Reservoir, south of Lemmon.

Petrified log, Petrified Wood Park, Lemmon.

are the Paleocene Ludlow formation. You can see very good outcrops of the lower half of the Ludlow formation in a short walk along the road on the south side of the reservoir. Fifty to 60 feet of brown sandstone, full of concretions of all sizes and shapes, separates two seams of lignite coal. You can see concretions and big chunks of sandstone in the riprap on the upper face of the dam.

Between Shadehill and Lemmon, the highway follows Flat Creek through clays and sands of the Cannonball formation. Lemmon stands on the Fort Union formation, which is above the Cannonball formation. It is the principal trade center for a large area of ranching and dry farming. Lemmon is best known to tourists for its Petrified Wood Park, built from hundreds of tons of petrified wood and concretions. The wood came from the Fort Union formation, the cannonball-like concretions from the Cannonball formation. A local company imports agate and other semiprecious stones from around the world and makes a line of jewelry and other ornamental items.

Hugh Glass was a guide for one of the fur-trapping parties that went up the various tributaries of the Missouri River to trap beaver in 1823. One day, separated from his companions, he stumbled upon a mother bear with cubs. The fight between them ended with the bear dead and Glass seriously mauled with a broken leg. His companions found him unconscious and nursed him for four days. Then, believing that he was going to die, they left with his food, gun, and matches. He later regained consciousness and crawled and rafted downstream, surviving on a diet of roots and berries, all the way to Fort Kiowa, 187 miles south on the Missouri River near Chamberlain.

Sturgis–North Dakota Line
121 miles

The rocks between Sturgis and the Moreau River crossing at Hoover are dark gray Pierre shale deposited in the late Cretaceous inland sea. Younger rocks north of Hoover are late Cretaceous Hell Creek clays and sands deposited on an arid alluvial plain after the Cretaceous sea had withdrawn. Streams brought the sediments in from the west. Their color is predominantly gray, with darker streaks of carbonaceous shale and impure lignite coal. The still-younger rocks in the Slim Buttes are White River beds similar to those in the Big Badlands.

Cretaceous formations of South Dakota.

Age	Lithology	Name, thickness, description
Upper Cretaceous		Hell Creek; 400–575 ft.; somber clays with concretions and sandstone ledges; some lignite and dinosaur remains
		Fox Hills; 50–200 ft.; shoreline sands and clays deposited by retreating Cretaceous sea
		Pierre; 1,200–1,500; shale, gray with several types of concretions; numerous persistent bentonite layers; divisible into zones on basis of lithology and fossils; noted for beautifully preserved cephalopods
		Niobrara; 175–225 ft.; Fort Hayes member, impure chalk; Smoky Hill member, chalk, weathers orange
		Carlile; 450–600 ft.; Sage Breaks member, gray shale with concretions; Turner member, gray shale with sandstone ledges; Pool Creek member, gray shale
		Greenhorn; 130–230 ft.; slabby limestone underlain by limy shale; thin shark tooth limestone at base; oily odor on fresh break
		Belle Fourche; 450–600 ft.; dark gray shale with few bentonites or concretions
Lower Cretaceous		Mowry; 175–225 ft.; light gray siliceous shale; thick Clay Spur bentonite at top
		Newcastle; 25–200 ft.; fine channel sandstone
		Skull Creek; 175–275 ft.; gray shale with few concretions or fossils
		Fall River; 75–275 ft.; fine sandstone
		Lakota; 200–600 ft.; Fuson member, variegated clay and sandstone; Chilson member, massive sandstone, thin clays

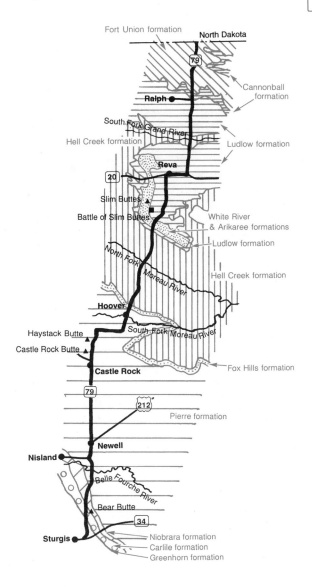

Sturgis to North Dakota line (121 miles).

Sturgis grew in a bend in the Red Valley where the 400-foot hogback ridge provided protection from the wind and Bear Butte Creek furnished water. Cliffs in the water gap through the hogback ridge expose more than 300 feet of early Cretaceous sandstone, all dipping gently down toward the northeast.

Bear Butte is the most easterly of a line of Tertiary igneous intrusions that stretch for 50 miles across the northern Black Hills to Devils

Bear Butte is an igneous intrusion into soft Cretaceous shales, now exhumed by erosion.

Tower and the Missouri Buttes in Crook County, Wyoming. It looms 1,200 feet above the prairie. When rising magma invaded the sedimentary rocks, it spread laterally between the layers, which were then horizontal, and it steeply tilted some as it rose through them. Look on the eastern side of the road for the vertical ledge of Minnelusa sandstone high on the western flank; the eastward dipping ridge that forms the eastern face of the butte is Madison limestone. The intrusive rock is a variety of rhyolite, about 51 million years old.

Next to the west flank of Bear Butte, an offshoot of molten magma raised a blister within the sedimentary layers, a small laccolith. Erosion has since stripped off the sedimentary layers domed over the intrusion, creating a bull's-eye pattern of outcrops, with the redbeds of the Spearfish formation at the center. The area within the bounding ridge of lower Cretaceous sandstone is called Circus Flats. The highway cuts across the east flank of the dome, passing rocks as old as the Spearfish formation.

A wildcat well that tested the dome structure for oil in 1921 found no oil, but it did uncork a huge artesian flow from the Mississippian Madison limestone at a depth of 800 feet. The water was diverted to Bear Butte Lake until 1987, when the well was plugged as a conservation measure.

Between Bear Butte and the Belle Fourche River, the highway crosses lower and middle Cretaceous formations that dip gently down to the northeast. The formations become successively younger to the northeast to the Pierre shale south of Newell. Outcrops are generally poor and overgrown with grass.

An oil test drilled at the base of Bear Butte found artesian water.

Just north of Bear Butte, the highway crosses the ruts of the old trail between Bismarck and Deadwood. It was an important stage and freight route from 1875 until the first railroad reached the northern Black Hills in 1890.

For several miles north and south of Newell, the highway crosses one of the early irrigation districts, started in 1910. Bedrock is mostly the bentonitic Pierre shale, much of which has become waterlogged and overloaded with mineral matter from the irrigation water.

Fox Hills sandstone caps the chain of low buttes north and west of Castle Rock Butte. North of the Moreau River at Hoover, the highway crosses a narrow outcrop of the Fox Hills formation too subtle to identify from the car window. The Fox Hills formation was laid down along a fluctuating shoreline as the late Cretaceous inland sea

Alkaline mineral salts from evaporation of groundwater in Butte County.

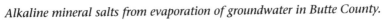

slowly retreated from the Great Plains region about 70 million years ago. It includes beach sands, barrier islands, coastal swamp deposits, and other types of shoreline features. North of Hoover, the depth to the Fox Hills sandstone aquifer gradually increases, but for many miles it is an easily accessible source of potable groundwater.

Somber gray clays of the Hell Creek formation are the bedrock between Hoover and the base of the Slim Buttes. The clays were laid down near the end of Cretaceous time after the inland sea had finally drained. Watch for the occasional small mud buttes. The heavy soil derived from these dark clays is better suited for sheep and cattle range than for dryland farming. Paleontologists have found any number of magnificent dinosaurs in the Hell Creek formation of the Dakotas and eastern Montana.

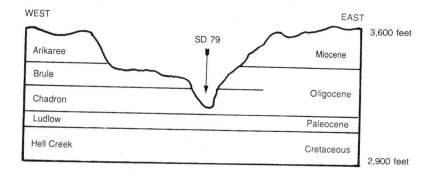

Profile through Summit Pass at the south end of Slim Buttes. The base of the section is at the elevation of the surrounding prairie.

Slim Buttes

The pale clays and sands exposed in the Slim Buttes are an erosional remnant of the White River beds that once covered the northern High Plains. Slim Buttes stand 600 to 700 feet above the surrounding prairie, giving some impression of the former depth of that cover and of the tremendous amount that was eroded, mostly during middle Miocene time. The next large remnant is in the Big Badlands of southwestern South Dakota.

SD 79 takes advantage of a low pass through a southeastern extension of the Slim Buttes. The much older Tongue River beds were probably deposited in this area during Paleocene time, about 60 million years ago. If so, they were entirely eroded during Eocene time, before deposition of the White River beds, which began about 40 million years ago.

North of SD 20, along the way to the North Dakota line, SD 79 crosses somber sediments of the Ludlow formation. They were deposited during Paleocene time, about 60 million years ago, a few million years after the dinosaurs vanished. The South Fork of the Grand River has cut through these deposits into the underlying late Cretaceous Hell Creek strata, which contain large numbers of dinosaur bones.

For 10 miles south of the state line, the Ludlow beds give way to sediments of the Cannonball formation. They were deposited during Paleocene time in very shallow seawater. The inland sea drained out of South Dakota as the Fox Hills formation was laid down in late Cretaceous time. But the inland sea still existed farther north and briefly returned to northwestern South Dakota during Paleocene time, when the area again sank slightly below sea level for the last time.

The marine clays of the Cannonball formation contain numerous round limestone concretions in the size range of citrus fruit and melons. They also contain the fossil remains of clams, snails, and other shellfish.

Lignite

From the Slim Buttes north, the sedimentary rocks of the Ludlow formation contain beds of lignite coal as much as 10 feet thick. The earliest white settlers started wagon mines before World War I, and they continued to produce coal through the bleak years of the Great Depression. A typical wagon mine started at an outcrop of lignite of suitable thickness and quality, usually close to a creek, then worked back by removing the overburden with horse-drawn scrapers. Blocks of the exposed lignite were loaded into wagons and hauled to ranches and schoolhouses. After World War II, use of local lignite declined rapidly in favor of fuel oil and liquefied petroleum gas.

THE BIG BADLANDS

After the Cretaceous inland sea retreated from western South Dakota and the dinosaurs met their terrible fate, Paleocene time began. That was when the Black Hills rose and began eroding to their present form. The dark gray Pierre shale east of the Black Hills eroded to a gently undulating plain.

The climate of Eocene time was very warm and very wet, certainly subtropical, perhaps even tropical. The iron pyrite in the shales oxidized to iron oxides in vivid shades of red and yellow to a depth

No wonder the early French explorers called them Mauvaises Terres a Traverser *or "bad lands to cross"!* —S.D. Department of Tourism

of many feet. This deeply weathered red soil was the kind you see in the modern tropics. You can still see it here and there in South Dakota, where it was buried under younger sediments. No sedimentary formations accumulated on land anywhere in the region until late Eocene time, presumably because the wet climate maintained vigorous streams that carried eroded sediments all the way to the ocean. Eocene volcanic rocks in the northern Rocky Mountains are full of petrified trees.

The Oligocene Landscape

Then, toward the end of Eocene time, the climate became arid or semiarid, converting South Dakota into a huge desert alluvial plain that sloped gently down to the east. About 600 feet of clays and sands buried the area that is now the Badlands. The Black Hills were buried to within 2,000 feet of their crest. These are the sediments that geologists call the White River formation, or the White River group of formations.

The fossil types within these sediments, combined with the character of the sediments, has enabled geologists to reconstruct the landscape of that time. It must have looked a lot like central Australia, or perhaps like some of the drier parts of East Africa.

Picture a broad and sparsely grassed plain with streams that are dry except when it rains. During storms, the streams overflow and drop much of their sediment load next to their channels. From time to time, those channels fill with sediment, and the stream finds a new course across the alluvial plain. The soil is fertile. Trees and shrubs thrive where they can find water. Elsewhere, the landscape is rather barren and very flat.

Fossil bones found within the gravelly stream channels are mostly those of animals that like the water, such as crocodiles, or of animals who accidentally drowned when they came to drink. In the nearby areas, the bones are primarily those of browsers. Some had short necks to feed on shrubs; others had long necks to reach the higher tree limbs. Most of the animals that lived out on the plains, away from the streams and the protection of the trees, were fast enough to outrun their enemies.

Some sedimentary layers contain abundant land snails and tortoises and the seeds of hackberry trees. Temporary ponds supported clams and snails, and mats of algae secreted lime to form lenses of limestone. The ponds attracted ducks and other birds, and occasionally delicate bird bones are found buried and preserved in the sediment; even fossil duck eggs have been uncovered.

All of that came to a halt in middle Miocene time, when the climate again became very wet and very warm. The streams flowed vigorously enough to erode the landscape and carry the sediment away instead of depositing it. In fact, they eroded a very large proportion of the White River group, leaving scattered remnants to tell the story.

Fossil eggs of a ducklike bird in the White River formation, about the size of pullet eggs. —S.D. Museum of Geology

Subdividing the White River Group

For nearly 150 years, geologists and paleontologists working in the Badlands have recognized certain distinctive and persistent layers that they can follow for long distances. They base these units primarily on the rocks, but the layers also contain distinctive fossils. In this book, we will lump those layers into two convenient and easily recognizable units: the Chadron formation in the lower part of the White River group, and the much thicker Brule formation in the upper part. Other designations are shown on the accompanying diagram; you can identify some of them at roadside stops.

Chadron Formation

The lowest formation of the White River group buried the colorful Eocene erosion surface in which the hills stood at least 150 feet above the streams. The maximum thickness of the Chadron formation is about 180 feet, so you see wide variation in its thickness between where it filled old valleys and where it covered hills.

The Chadron formation consists mostly of pale greenish clay that contains enough bentonite to make it expand and become very slick and sticky when wet. When it dries, it shrinks to irregular fragments that suggest popcorn. The formation erodes to gentle slopes and low haystack buttes. The clay is so plastic that where it caps hills, it tends to creep down over the strata beneath.

Haystack weathering of the Chadron formation. Cliffs behind are of the overlying Brule formation. —J. E. Martin

Thin lenses of limestone in the Chadron formation record shallow ponds in which the water must have been alkaline and saturated with calcium carbonate. The limestones contain fossils of algae, snail shells, and clams. They look like the kind of limestones that accumulate in shallow ponds in arid regions, where drainage of flat surfaces tends to be very poor.

The Chadron clays also contain the fossil bones of animals such as alligators, which probably lurked in the streams, and big tortoises that probably avoided the water. Mammals abounded, including the gigantic titanotherium, which must have resembled a rhinoceros with a severe glandular problem.

Brule Formation

Sediments in the Brule formation are a banded mixture of water-laid clays, fine windblown sands and silt, and volcanic ash. Its maximum thickness is about 450 feet. The Brule formation weathers to steep pinnacles and to cliffs that suggest flights of stairs. These erosional flights of fancy are the fairy castles of the Big Badlands.

Most of the layers in the Brule formation are tan with a generally pinkish cast, but in the lower part they tend to have color bands of dark red, purple, and gray. The colors are particularly vivid after a rainstorm. Small limy nodules or concretions are characteristic of

View from Pinnacles Overlook. Rocky Ford ash is in the foreground, and banded beds of underlying Brule formation are in the background. —R. W. Wilson

some layers; some of those nodules may be remnants of desert caliche soils. Lenses of strongly cemented channel sands and gravel cap and protect small clay buttes.

The Brule formation contains one of the richest troves of mammal fossils in the world. They include, so far, more than 150 genera, and range in size from rodents smaller than a mouse to a rhinoceros. Then, as now, carnivores ate the plant eaters. Most modern groups of mammals have ancestors in that fauna, including horses and camels, which originated in North America. But, a good many of the animals that left their bones in the Brule formation have no modern descendants. During the 5 million years in which the Brule formation was laid down, some of the animal species died out, some evolved into new forms, and some remained virtually unchanged.

Toward the close of Brule time, increasing amounts of volcanic ash drifted into western South Dakota on the wind. Many geologists think it erupted in the Western Cascades of Oregon. Watch for the dazzling white layer of Rocky Ford ash that overlies the less pure ash layers in the uppermost Brule formation. Its innocent sparkle tells of an extremely large and violent rhyolite eruption somewhere far to the west.

Younger Strata

Then, after a few million years, the wet period ended, the streams lost most of their flow, and deposition of sediment resumed. The new series of sediments accumulated intermittently during late Miocene and Pliocene time until the coming of the great ice ages about 2 million years ago.

Only small patches of those younger sedimentary deposits remain in the Big Badlands. In the southwestern part of the state, and in adjacent states, hundreds of feet of younger sediments cover the White River group. These are mostly pink clays and sandstones brought in by wind and water during Miocene time. Some of the sandstones are solid enough to make fairly prominent cliffs.

In some places, the late Miocene sandstones overlie the White River group; elsewhere they fill valleys cut into the older sediments. Collectively, the late Miocene and Pliocene sediments are known as the Ogallala group. Farther west, on the High Plains east of the Rocky Mountains, the Pliocene sediments are much coarser and form the great Ogallala aquifer. What we today call the High Plains are remnants of the old desert alluvial plain surface that collected sediment of the Ogallala group during late Miocene and Pliocene time.

The Big Badlands, as we see them today, were eroded very recently. The most spectacular landscapes are next to the White River. That river gives its name to the several hundred feet of clays known as the

White River group. When we speak of the White River group, we refer to the pallid beds laid down during late Eocene and Oligocene time, regardless whether they are in the Big Badlands or in isolated buttes scattered across the prairie. Similarly, the fossil remains of the animals that thrived during that time are known, everywhere, as the White River fauna.

Passes

A pass, in the local sense, is a place where it was possible to drive a wagon or a herd of cattle up or down a badlands cliff. Passes are typically the result of landslides that reduce the steepness of an erosional wall. Those slides generally interrupt the local drainage. Damming of even small gullies impounds enough water to saturate the soil, thus causing further slumping. The extra moisture, and the shelter along the badlands walls, offer a toehold for shrubs and trees. Those isolated patches of green vegetation at the passes shelter many species of mammals, reptiles, and birds.

At either Cedar Pass on the east, or the Pinnacles on the west, the highway descends through several hundred feet of distinctly layered sands and clays, the White River beds. On the basis of rock type, color, concretions, and sandstone layers, geologists divide the rock sequences into several units. These also contain distinctive assemblages of fossils. Geologists trace the rock units with their fossils throughout the Big Badlands.

SD 240

The Badlands Loop, Cactus Flat to Wall
39 miles

This scenic route winds through the Big Badlands. It connects with I-90 at Cactus Flat, Exit 131, on the east and at Wall, Exit 110, on the west. It offers a trade of 39 miles with lots of spectacular scenery for 22 miles of monotonous interstate driving.

The scenic area lies a few miles south of the interstate, so you face a few miles of flat, upland driving at either end. The spectacular part of the route lies within the Badlands National Park. In return for a modest admission charge, you get a small brochure describing the various pulloffs. These offer scenic views, instructive displays, or short nature trails. The road in the park follows the Badlands Wall, the escarpment that separates the dissected lowland along the White

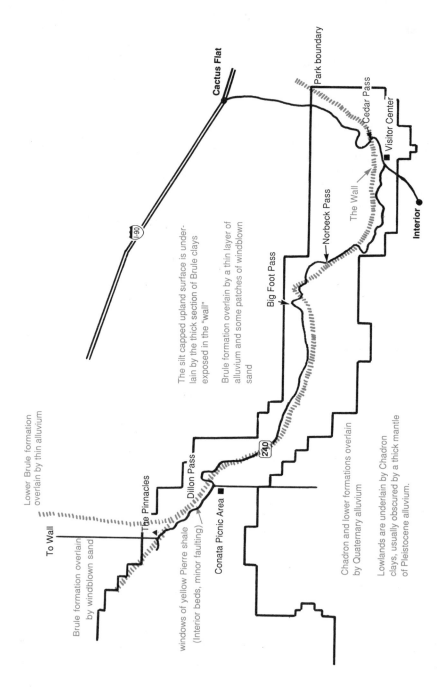

Cactus Flat to Wall (39 miles).

Cactus Flat

Park boundary

Cedar Pass

Visitor Center

The Wall

Norbeck Pass

Big Foot Pass

Interior

The silt capped upland surface is underlain by the thick section of Brule clays
exposed in the "wall"

Brule formation overlain by a thin layer of
alluvium and some patches of windblown
sand

Dillon Pass

240

Conata Picnic Area

Chadron and lower formations overlain
by Quaternary alluvium

Lowlands are underlain by Chadron
clays, usually obscured by a thick mantle
of Pleistocene alluvium.

Lower Brule formation
overlain by thin alluvium

The Pinnacles

To Wall

Brule formation overlain
by windblown sand

windows of yellow Pierre shale
(Interior beds, minor faulting)

Looking south from the Badlands Wall. Banded beds are Brule formation. —J. E. Martin

River to the south from the uneroded upland to the north. In some places, the highway hugs the base of the cliff; in others, it follows the rim.

Between Cactus Flat and Cedar Pass, the upland surface is underlain by Brule formation, but you can hardly see it. Several feet of alluvial silt and a few patches of windblown sand overgrown in grass cover the surface.

Numerous exhibits in the visitor center at the park headquarters offer an introduction to the animal and plant life, rocks, and cultural history. An excellent assortment of books and maps is on sale. As you stand outside the center, try to follow individual rock layers from butte to butte.

From the park headquarters to Norbeck Pass, the road follows the base of the Badlands Wall. The road is built upon a mantle of alluvium eroded from the wall and spread over the lowlands. Low buttes with a cover of sod are remnants of the surface that existed before recent erosion started washing away the easily erodible material.

This stretch of the Badlands Wall consists of clays and sands of the Brule formation. They contain a zone of very small faults that trend west or northwest, but they are hard to see. Watch for the landslide terrain at Norbeck Pass.

Siltstone concretions weathering out of the Brule clay at the Window pulloff at Cedar Pass.

Sod-covered tables are remnants of alluvium.

Butte of Brule formation, north side of highway at the visitor center.

The route between Norbeck Pass and Dillon Pass lies along the upland rim. An alluvial soil supporting a heavy grass cover and suitable for dryland farming covers the upland surface. The sod protects the upland surface from badlands erosion. Several pulloffs provide good views for photographs.

Along a short interval at Big Foot Pass, the highway traverses the full section of the Brule formation. Watch for the transition between the pinkish tans of the silty Brule formation and the pale greenish clays of the underlying Chadron formation.

At Dillon Pass, the road goes all the way through the greenish clays of the Chadron formation and into the late Cretaceous Pierre shale beneath. Watch for the orange and yellow clays of the weathered soil zone at the top of the Pierre formation. They are soils developed on Pierre shale in early and middle Eocene time and buried under the Chadron formation during late Eocene time. Watch along the road for large limestone concretions, typical of the Pierre shale. At one pulloff, you can climb a small butte and see a fault that displaces the Chadron and Brule beds about 50 feet.

This short segment of road between Dillon Pass and the Pinnacles crosses the entire section of White River beds, all the way from the weathered Pierre shale at the base to the layer of sparkling white volcanic ash at the top. Above that is the Sharps formation.

Upper Brule clay at the Pinnacles, dried out and cracked. Sand and clay filtered into the crack to form a sediment-filled dike. It appears as a steep, dark line cutting the right-hand pinnacle.

The state highway department designed and built this short section of highway in the 1920s to introduce tourists to the Big Badlands. The department graded it with horse-drawn equipment and a lot of manpower. From the upland at the top of the Pinnacles, a good gravel road, Pennington County 590, circles the Sage Creek Wilderness Area, which supports a small herd of reintroduced buffalo.

At the Pinnacles, the Badlands Wall swings due north. A layer of windblown sand and silt, which collected during the great ice ages, covers most of the upland surface. The highway to the town of Wall lies on the upland surface roughly 0.5 mile west of the Badlands Wall. No overlooks exist.

The Black Hills

The Black Hills were one of the last areas in the United States to be examined by trained scientists. George Armstrong Custer's 1874 expedition to the Black Hills was followed in 1875 by a smaller but scientifically trained group known as the Jenney-Newton expedition. The two leaders were geologists. In one brief summer, they compiled a great deal of information about the stratigraphy and location of mineral occurrences, but it took a while to assimilate all their field data. Newton returned to the Black Hills the following summer to verify some fine points of the geology, but died of typhoid fever in Deadwood. Petty jealousies in Washington delayed issuance of the final report and maps until 1880, so they were not immediately available as a guide for prospectors. Jenney's professional career took him away from the Black Hills, but he returned briefly in 1893 to serve as dean of the School of Mines in Rapid City.

Bird's-eye view of the Black Hills. —Jenney-Newton survey, 1875

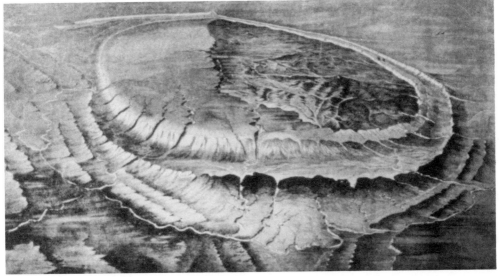

The Black Hills first appear to the approaching traveler as a low, dark mound rising from the seemingly endless prairie. As you approach, the outline of a modest mountain range develops, standing 4,000 feet above the rangeland. The dark color, especially evident in winter, is due to a heavy cover of pine and spruce trees. Still closer, a sharp delineation between the rolling prairie and the mountains becomes evident.

Geology of the Black Hills.

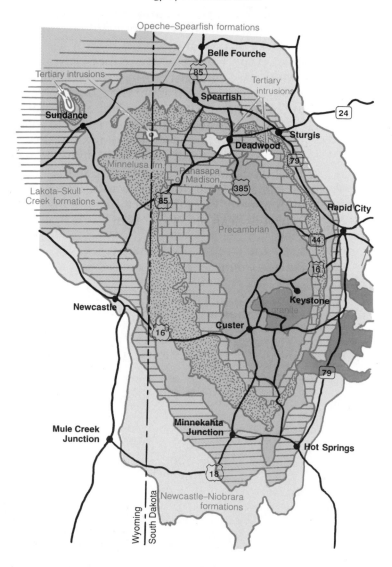

The Black Hills are the only part of South Dakota where you can see the older sedimentary formations. Some of these formations are very important to the economy of the state because they supply water to deep wells; without them, much of western South Dakota would be unpopulated. But these formations are buried everywhere except in the Black Hills.

Paleozoic Rocks

The ancestors of all the animals now living, including ourselves, appear as fossils first in sedimentary rocks deposited about 600 million years ago. Their advent marks the beginning of Cambrian time. Simultaneously, sea level began to rise, and seawater flooded across all of the continents. The rising sea finally flooded across western South Dakota late in Cambrian time.

The landscape was then a fairly low and monotonous plain with a few topographic eminences, such as the long ridge of Sioux quartzite. The shallow sea deposited sand, now the Deadwood sandstone, as it advanced. In the Black Hills—the only part of South Dakota where Cambrian rocks are exposed—the Deadwood sandstone is about 400 feet thick in Spearfish Canyon, less than 50 feet thick near Wind Cave. Wells drilled a few miles south of Wind Cave reveal no Cambrian rocks, apparently because the wells are south of the old shoreline. Wells drilled in the northwestern corner of the state penetrate more than 700 feet of Cambrian Deadwood formation.

The lower part of the Deadwood formation is beach sands and gravels. As the shoreline moved east and the water deepened in the west, muds and thin layers of limestone were deposited. Tracks, trails, and borings of worms and other primitive forms abound on the bedding surfaces, evidently showing that they were exposed at low tide.

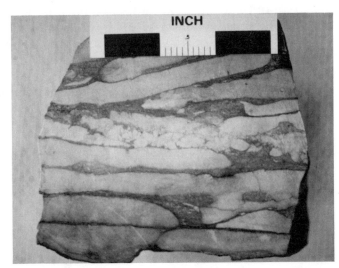

Intraformational conglomerate from the Deadwood formation. —Phil Bjork, S.D. Museum of Geology

A trilobite from Cambrian rocks. This well-preserved specimen, about 1.5 inches long, is from Nevada. —Phil Bjork, S.D. Museum of Geology

Beach sands were again deposited as the sea shoaled and the shoreline retreated west at the end of Cambrian time, and now those sands make the upper few feet of the Deadwood formation. You can see it all in the town of Deadwood. In fresh exposures, the sandstones are nearly white and the limestones and shales are gray. The latter weather to a distinctive reddish brown color.

The Deadwood formation also contains intraformational conglomerates, in which fragments of limestone are embedded in a limy shale. Evidently, thin layers of soft limestone accumulated on the seafloor, then waves broke them up, rolled the pieces about, then incorporated them in deposits of limy mud. These rocks exist in the middle part of the Deadwood formation, but they are rarely easy to see.

The most common fossils in Cambrian rocks are trilobites, extinct relatives of the crabs and shrimps. These very complex creatures abound in most Cambrian deposits, but they are hard to find in the Deadwood formation. Trilobites molted periodically as they grew, and most of the fossils are fragments of molted skins. The Deadwood sea did not leave South Dakota until very early in Ordovician time, for the highest trilobites in the rock section along Spearfish Canyon are Ordovician.

After the Cambrian period came Ordovician time. Ordovician rocks now underlie northwestern South Dakota. You can see them at the surface only in the northern Black Hills. They include none that look like anything that would accumulate along a shoreline, so the

Dicellomus, a primitive brachiopod from beach sand in the Deadwood formation of the southern Black Hills, where it is common. Specimen is 4 inches across —Phil Bjork, S.D. Museum of Geology

original blanket of sediments was probably much larger and has eroded away. The oldest Ordovician formation is the grayish green Winnipeg shale, which appears in the Black Hills and near Winnipeg, Canada. It is 40 to 70 feet thick. Some geologists call it the Icebox shale for good exposures along US 85 in Icebox Canyon, southwest of Lead.

Above the Winnipeg shale are about 30 feet of Roughlock siltstone, which grades upward into the Whitewood dolomite, named for outcrops along Whitewood Creek below Deadwood. The Whitewood formation consists of about 60 feet of coarsely crystalline dolomite, which is light gray when fresh, but weathers to striking orange and brown on outcrops. The Whitewood dolomite extends for great distances to the north and west, where it is called the Bighorn dolomite in Wyoming and Montana.

Fossils are moderately abundant in the Ordovician rocks. The Winnipeg shale contains only microscopic fossils, tiny teeth of marine worms and odd plates known as conodonts, which look like little jawbones with rows of teeth. The Roughlock siltstone contains very few fossils, but the Whitewood dolomite contains fossils of many varieties of brachiopods, corals, marine snails, and large, straight conical species of Orthoceras related to the present-day pearly nautilus.

After Ordovician time was Silurian time, followed by Devonian time. Silurian and Devonian rocks exist in northwestern South Dakota, but they do not reach the surface, not even in the northern

*Maclurites, a marine snail
from the Ordovician
Whitewood dolomite.
Specimen is 3.5 inches across.*
—Phil Bjork, S.D. Museum of Geology

Black Hills. Deep wells penetrate several hundred feet of Devonian dolomites, shales, and sandstones, then pass into several hundred feet of white Silurian limestone. The thickness of the two combined is about 900 feet in northwestern Perkins County. They thicken northward into North Dakota, where they produce oil.

Then came Mississippian time, a period in which South Dakota and many other parts of the world acquired large amounts of limestone. During early Mississippian time, the shallow inland sea again flooded a large part of North America, laying down enormous deposits of limestone. In South Dakota, the Pahasapa, or Madison as many Rocky Mountain geologists prefer, limestone exists only in the western part of the state and is exposed only in the Black Hills. The formation is called the Madison limestone for outcrops along the Madison River in Montana.

The advancing sea of Mississippian time flooded an old erosion surface covered with red soil, some of which was incorporated into the lowest part of the limestone. This basal section of pink limestone is the Englewood formation. It is overlain by several hundred feet of pale gray rock, the Madison limestone. The contact between the Englewood and Madison formations is where the color changes from pink to gray.

In northwestern South Dakota, the Englewood and Madison limestones thicken from zero just south of the Black Hills to over 1,200 feet at the western end of the state. Some of the change in thickness is an original depositional feature, some is due to erosion. In the

Black Hills, the thickness ranges from a little over 300 feet in the south to 600 feet in the walls of Spearfish Canyon.

Invertebrate fossils exist in the Englewood and Madison formations, but getting them out of the rock intact is nearly impossible. They include corals, crinoids, and an assortment of brachiopods. Some of the brachiopods are hollow and lined with pretty crystals of white calcite.

After the Madison limestone was laid down, the land rose above sea level, exposing huge areas of newly deposited limestone to weathering and erosion. Nearly 50 million years passed before the sea again flooded the region during early Pennsylvanian time. During this interval, rainwater, made slightly acid during its passage through the air, seeped slowly down through the limestone layers, dissolving caverns in the rock. As the caverns collapsed, many of them broke through to the surface, opening sinkholes. This type of cave and sinkhole topography is called karst, after a region in eastern Italy, western Slovenia, and western Croatia.

Pennsylvanian time followed Mississippian time. During the early part of Pennsylvanian time, shallow seawater again flooded across South Dakota, this time from the south. It reworked a red soil that had accumulated on the Mississippian limestone terrain, washing it off the high spots and filling the sinkholes. Alternating beds of fine sand and dolomite were laid down on top of the red soil. The sea retreated for a short period, then returned to deposit thin layers of sandstone, shale, dolomite, and gypsum as the lower Minnelusa formation.

Sometime near the end of Pennsylvanian time, the sea retreated long enough for another red soil to form. When the sea returned, the red soil was reworked into a layer of red rock, which drillers call the red marker bed. It enables them to predict how much farther they have to drill to reach some local oil sands, or to enter the top of the Madison limestone. The upper Minnelusa formation was laid down when deposition resumed in Permian time. It contains more gypsum and very fine sand that may have blown in on the wind. The top of the formation contains dune sands that appear to have blown along the coast.

After the Minnelusa formation was deposited, the Opeche formation was deposited on top of it during early Permian time. It consists mostly of redbeds, red sandstone, and mudstone laid down above sea level and interlayered with occasional beds of white gypsum. It all looks like stuff deposited in a desert.

A shallow sea again flooded western South Dakota during the middle part of Permian time and deposited the Minnekahta limestone. It is an extraordinary rock, some 40 feet of remarkably pure limestone in very thin layers. It contains very few fossils, perhaps because the water was too salty for animals to survive.

Then, in late Permian and early Triassic time, the region again rose above sea level and more redbeds were laid down on dryland. These are called the Spearfish formation in South Dakota, the Chugwater formation in Wyoming and Montana. This tremendous blanket of red sandstone and mudstone thickens westward toward its probable source in Utah. The redbeds exist under the western third of South Dakota to a thickness of more than 700 feet along the Wyoming line.

The end of Permian time, which was also the end of Paleozoic time, came while the redbeds of the Spearfish formation were accumulating. That end came catastrophically with the rapid extinction of more than 90 percent of all the animal species then living.

Mesozoic Rocks

Mesozoic time dawned on an earth nearly devoid of animals. The Spearfish formation was still accumulating, so it must bridge the boundary between Paleozoic and Mesozoic time. It is not fossiliferous, so it opens no window to the events that destroyed most of the

An oyster shell bed in the Niobrara formation. —Phil Bjork, S.D. Museum of Geology

earth's animals at the end of Permian time. Neither does it tell us anything about how the survivors evolved and spread across the earth as Mesozoic time began with the Triassic period.

From middle Triassic to middle Jurassic time, some 50 million years, South Dakota was above sea level and eroding. Strata of Triassic and older age were progressively beveled toward the east. The 700 feet of redbeds at the town of Spearfish thins to zero near Wall, and the Minnelusa and Madison formations successively thin out and the lower Paleozoic rocks form the old erosion surface.

During much of Jurassic time, a highland extended from southern California to Idaho. During late Jurassic time, shallow seawater flooded eastward into the Black Hills area. Several hundred feet of marine sandstone and shales, the Sundance formation, accumulated. The Sundance sea retreated very briefly about the middle of that time interval, and a thin streak of redbeds now separates the Sundance formation of South Dakota into lower and upper members.

The Sundance formation is probably the most fossiliferous marine deposit in South Dakota. Clams and oysters abounded on the sandy bottoms, and cephalopods swam in the water above the muddy bottoms. Belemnites, relatives of squid, thrived everywhere; their remains, which are about the size, shape, and color of cigar butts, abound in all kinds of rocks. The Sundance formation ranges from less than 300 feet to nearly 450 feet in thickness. Most of the good outcrops are high on the inner flanks of the Cretaceous hogback around the Black Hills.

Above the Sundance formation on the southern and eastern flanks of the Black Hills lies a thick sandstone known as the Unkpapa formation. In texture, it so closely resembles the Sundance sandstones that it either was derived from the same source or consists of sand eroded from the Sundance formation. The many small faults that

Belemnites about 3 inches long.
—Phil Bjork, S.D. Museum of Geology

offset its thin layers appear to have formed as the sand compacted before it was cemented into solid rock. At Rapid City, the Unkpapa sandstone is dazzling white; north of Buffalo Gap and near Hot Springs it is lavender; at Calico Canyon, just south of Buffalo Gap, it is finely banded with thin layers of yellow, red, and white. All the color is due to traces of iron and manganese oxides, which were probably introduced into the sandstone by circulating groundwater. The formation contains no fossils.

Overlying the Unkpapa sandstone, or the Sundance shales where the Unkpapa sandstone is missing, is the grayish green Morrison shale. It accumulated along the eastern edge of the area that held the Sundance sea. The Morrison formation is widely noted for its dinosaur bones; Dinosaur National Monument in Utah is an excellent place to see them. In the Black Hills, some of the Morrison formation was eroded before Cretaceous rocks were laid down on top of it, so the remaining thickness ranges from zero to 220 feet.

In early Cretaceous time, about 100 million years ago, the Rocky Mountains were beginning to rise from western Canada to western

Small faults offset the thin layers in Unkpapa sandstone.
—S.D. Museum of Geology

Mexico. A broad trough began to subside east of those new mountains, most rapidly near its western margin. Meanwhile, the shallow inland sea again flooded in from both north and south to make a seaway that extended from the arctic to the Gulf of Mexico and eastward to the western edges of Minnesota, Iowa, and Missouri. Streams dumped their sediments into it, and waves and currents distributed them along the shorelines and across the seafloor. By the end of Cretaceous time, only a few hundred feet of sediment had accumulated along the eastern side of the seaway. The deposits thicken westward to more than 4,000 feet in northwestern South Dakota, much more farther west.

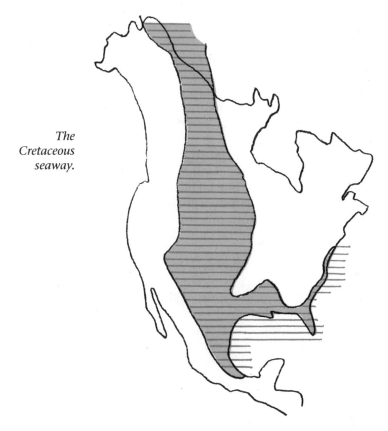

The Cretaceous seaway.

As Cretaceous time neared its end, about 70 million years ago, the trough was filled with sediment and the sea slowly retreated to the north and south. Streams continued to spread mud and sand over the area as it emerged above sea level to become a broad coastal plain. Dinosaurs returned to wander where the seas had been, until their extinction at the very end of Cretaceous time, almost exactly 65 million years ago.

Sediments deposited on that coastal plain during late Cretaceous time are called the Hell Creek formation in South Dakota and eastern Montana. Most are muds and sands, mostly in somber shades of gray. They weather to gumbo badlands, sticky and slippery in the rain.

Studies of old stream channels preserved in the Hell Creek formation suggests that streams then flowed east across the present site of the Black Hills. Evidently, the Black Hills had not yet started to rise. If so, much of western South Dakota was once covered with a thick blanket of Hell Creek clays and sands. By the end of the Cretaceous period, approximately 6,500 to 7,000 feet of Paleozoic and Mesozoic sediments lay upon the old Precambrian surface.

Rise of the Black Hills

Millions of years before the Black Hills rose, western South Dakota was part of a vast plain standing near sea level. At times, it lay below sea level and received successive layers of limestone, sandstone, and shale. Between invasions of the shallow seas, it stood slightly above sea level. During some of those times, it received additional deposits of sand and gravel, at other times it eroded. Altogether, about 7,500 feet of flat strata accumulated during this long interval of deposition and occasional erosion.

The oval dome of the Black Hills lies astride the western border of South Dakota and extends west into Wyoming and north into extreme southeastern Montana. It probably started to rise about 62 million years ago at about the same time as the mountains of central Wyoming and central Montana, and it was as high as it would ever get by about 48 million years ago.

The rise of the Black Hills raised the ancient metamorphic rocks and granites to the crest of Harney Peak, 7,240 feet above sea level. If we were to restore the 7,500 feet or so of sedimentary rocks that were eroded to expose the Harney Peak granite, we would have a mountain rising out of the prairie to an elevation of nearly 15,000 feet. But that sedimentary cover probably eroded as the Black Hills rose, so the mountains were never that high.

By roughly 37 million years ago, the Black Hills had much of their present size and shape. Hard rocks formed the ridges and plateaus, softer rocks became valleys and parks. From the air, the Black Hills resemble a huge, oval target with the oldest rocks in the bull's-eye and successive rings of younger rocks around it.

Erosion has entirely removed the younger rocks over an area 50 miles long from north to south and 20 to 25 miles wide from east to west. Most of the rocks exposed in this area are metamorphic rocks that began as muddy sandstones and were then squashed and re-crystallized. Now they are slates and schists in drab shades of gray. Large masses of molten granite magma invaded them while they were hot.

Two small areas feature small outcrops of very old granite that was emplaced about 2.5 billion years ago. One peeks out from under much younger rocks along Little Elk Creek west of Tilford; the other lies under Paleozoic rocks near the boy scout camp at Bear Mountain northwest of Custer. Both granites intrude metamorphic rocks that are older than the granites, and each granite has been metamorphosed into gneiss.

Very old sedimentary rocks near Nemo include two separate sequences of banded-iron formation, which consist of alternating thin layers of dark gray to black hematite, iron oxide, and nearly white recrystallized chert. The sequences closely resemble some of the ore in the Minnesota iron ranges, which for many years was the source of much of American steel. In the time between deposition of the two iron formations, a thick blanket of sand accumulated, later to become sandstone, and when metamorphosed, became the Box Elder quartzite. While the sandstone still lay flat, a sheet of molten gabbro about 3,000 feet thick intruded between the layers of sandstone. Age dates show that the gabbro is about 2.15 billion years old, so the sandstone (now quartzite) is still older. Vertical faults later broke the Box Elder quartzite, its gabbro sill, and the banded-iron formations to make basins, which accumulated stream boulders and gravels derived from the older iron formations and quartzite. Then all the rocks were folded so that the bottoms of the basins tilted past 90 degrees. It all makes an extremely complex mess.

Then the Harney Peak granite and its associated pegmatites was emplaced at a depth of possibly 12 to 15 miles. The granite domed up so that the outer parts cooled or crystallized at less depth, about 8 miles.

Much younger igneous rocks formed as the Black Hills rose. A belt of igneous activity developed during Eocene time across the northern end of the bulge from the Devils Tower and Missouri Buttes on the Wyoming side east 70 miles to Bear Butte. Most of the activity happened between 58 and 50 million years ago.

The igneous rocks of Eocene time in the northern Black Hills look quite different from the old granite of Precambrian time, which is so conspicuous in the southern Black Hills. The old granite is coarsely granular with obvious crystals of feldspar, mica, and quartz; the younger igneous rock is generally very finely crystalline, lacking visible quartz. Some of the younger rocks, with scattered coarse crystals of feldspar embedded in a very fine matrix, have a porphyritic texture.

Crystals of potassium feldspar up to 2 inches long, which have weathered free from a Tertiary porphyry.

Where they invaded flat-lying sedimentary formations, the molten magmas of Eocene time typically followed layers of softer rock, raising those above into domes a mile or more across and as much as 1,000 feet thick. The resulting blisters of igneous rock are called laccoliths. Some are isolated; others overlap one another in clusters. Where the magmas of Eocene time invaded the ancient Precambrian metamorphic rocks, they generally filled fractures to form tabular masses called dikes that typically stand at steep angles. In some cases they make small lenses a few feet across.

The igneous rocks of Eocene time are chemically peculiar in containing abnormally high concentrations of sodium and potassium. They form a wide variety of igneous rock types, possibly because they originated from at least two masses of molten magma within the upper part of the earth's mantle, possibly as much as 30 miles below the surface. And the magma may have melted and assimilated some of the rock through which it passed on its way to the surface. In at least one case the molten igneous rocks erupted through a volcano.

Heat from the molten magmas of Eocene time drove circulating hot water and steam, which deposited metallic ores both in the igneous rocks and in the older rocks that enclosed them. Except for the Homestake Mine, nearly all the old mines in the northern Black Hills worked ore deposits that those solutions emplaced in the Deadwood formation. The ores concentrated in two zones of sandstone and dolomite, one near the base of the formation, the other near the top. Most of the ores were mainly gold, some were mainly lead and silver, and a few were mainly tungsten.

The question of when the High Plains rose from near sea level to their present elevation has no clear answer. In western South Dakota, the plains rose more than 2,000 feet, in eastern South Dakota only a few hundred feet. Some geologists believe that happened during Eocene time, about when the Black Hills were bulging through the plains. Others believe it happened during Pliocene time, within the last few million years. None can support their view with compelling evidence.

The Anomalous Location of the Black Hills

When you look at a map of the central United States, you may wonder why this small mountain range rose so far east of the Rocky Mountains with attendant igneous activity. The geologic structure of the Black Hills dome and its age strongly suggest that the Black Hills rose in response to the same forces that raised the isolated mountain ranges of central Montana and Wyoming at the same time. In a way that no one clearly understands, those forces were probably related to the collision between North America and the floor of the Pacific Ocean, which was in progress when the Black Hills rose and continues today.

The Black Hills Landscape

Most of the erosion that shaped the modern landscape of the Black Hills happened during late Paleocene and early Eocene time, when the climate was very wet and very warm. Buried early Eocene soils are red and yellow, like the modern spoils of the humid tropics, and they contain the same minerals.

Then, near the end of Eocene time, the climate dried and the streams lost much of their flow. Streams no longer carried the debris from the Black Hills to the ocean. Instead, it accumulated on land. The new sediments were mostly very pale, partly because they included tremendous volumes of volcanic ash that blew in from far to the west. The coarser gravels that accumulated in stream channels came from local sources and represent all the types of rocks in the Black Hills. Geologists call these sediments the White River group. This burial continued until a thick blanket of young sediments left less than 2,500 feet of ancient rocks standing above a broad desert alluvial plain.

Then, in middle Miocene time, the climate again became wet and warm. Much of the blanket of sediment that accumulated during

Oligocene and early Miocene time was eroded. Patches remain up to an altitude of 5,200 feet. Thick blankets also remain on the stream divides east of the Black Hills.

After that episode of erosion, the climate again dried, and the streams again lost much of their flow. A new generation of sediments buried the landscape that had eroded when the climate was wet, and they spread across the land in a desert alluvial plain that stretched flat across all of western South Dakota. That generation of sediments accumulated through late Miocene and Pliocene time.

The climate again became wet enough to maintain stream flow about 2 million years ago, when the great ice ages began. Now we see the modern streams eroding the sedimentary cover off the Black Hills, exhuming landscapes that have been buried for a very long time. To some degree, those landscapes may have been eroded during middle Miocene time. But to a much larger degree they were eroded during early Eocene time. Most of the larger elements of the Black Hills landscape, the major ridges and broad valleys, were eroded to essentially their present form during Eocene time, about 50 million years ago.

Most of the major streams that leave the Black Hills start as springs in the Paleozoic limestone plateau on the western side of the Black Hills. They flow in deeply incised canyons across the core of Precambrian basement rocks and the narrow band of Paleozoic limestones. Then they cross the Red Valley and flow through canyons and water gaps in the hogback ridge of Cretaceous sandstones that encircles the Black Hills.

In general, the streams gain slightly in flow as they cross the dense Precambrian schists. Then they lose some or all of their water to caverns in the Paleozoic limestones. Unless they receive additional water from springs in the Red Valley, they pass through the hogback and start across the prairie with minimal flows. All the Black Hills streams are short. They join either the Cheyenne River, which swings around the southern end of the Black Hills or the Belle Fourche River, which encircles the northern end.

The Cretaceous Hogback

The resistant ridge of Cretaceous sandstones, mostly of the Lakota formation, that completely encircles the Black Hills stands 300 to 400 feet above the surrounding prairie and is the boundary between the Black Hills and prairie. Using the ridge as the outer limit, the Black Hills appear as a dome uplift that extends more than 120 miles from north to south and between 50 and 60 miles from east to west. More subtle ridges of limestone, chalk, and sandstone outside the

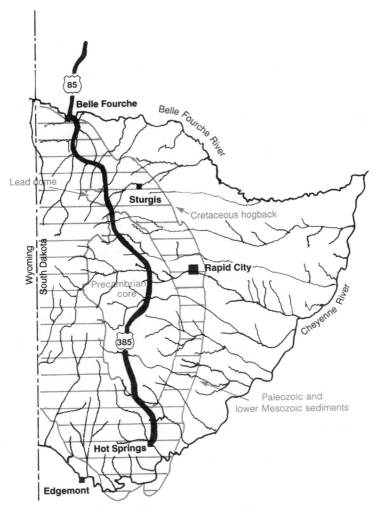

Radial drainage pattern of the Black Hills.

big hogback show that geologic boundaries of the dome are considerably larger.

The outer surface of the Cretaceous hogback is a slope of Fall River sandstone dotted with small pine trees. On the eastern margin of the Black Hills, at Rapid City, the sedimentary rock may locally dip down as much as 14 to 21 degrees. On the northern and southern ends, the dip is much less, but on the far western edge in Wyoming, the westward dip may reach 30 degrees locally.

The Fuson shale below the Cretaceous Fall River sandstone has eroded to form a break in the slope of the hogback; the massive sandstones of the lower unit of the Lakota formation form the true crest of the ridge. If the sandstones were projected over the Black Hills, they would lie 2,200 feet above Harney Peak.

The inner slope of the hogback is steep. It is generally capped by a cliff of Lakota sandstone, underlain by easily erodible Morrison shale and Unkpapa sandstone. The lower half of the slope is on the Sundance formation, which is predominantly a soft grayish green shale, but includes persistent ledges of sandstone at top and bottom. Some thin layers of limestone, basically a mass of small oyster shells, often crop out midway up the slope.

The Red Valley

The Red Valley extends entirely around the Black Hills just inside the high ring of the Cretaceous hogback. It is a broad valley eroded in the soft greenish shales of the Sundance formation and the red siltstones and streaky white gypsum of the Spearfish formation. Early residents called the Red Valley the racetrack.

During a first stage of erosion, the valley was excavated to a depth of at least 300 feet below the crest of the Cretaceous hogback. Erosion continued until the valley floor was a wide flat surface with a cover of gravel. Erosion is now deepening the valley, but numerous remnants of the old terraces are visible next to I-90 between Rapid City and Sturgis.

The Red Valley is narrow where the rocks are steeply tilted, wide where the dips are low. Its deeper parts offer an inviting route, used first by ox trains hauling freight and then by stage lines bringing passengers to the mines of the northern Black Hills. The railroads and then the highways followed that route.

The Red Valley is long, but has no permanent stream. Short streams that flow during wet weather carry sediment to the few main streams that cross the Red Valley, which leave the Black Hills through gaps in the Cretaceous hogback.

The inner side of the Red Valley is on a long and gentle slope of the very resistant Minnekahta limestone. Although only 30 to 35 feet thick and underlain by 100 feet of soft red clays in the Opeche formation, the limestone resists erosion and forms a sloping pavement ranging from a quarter of a mile wide where the rocks dip steeply toward the valley to 2 miles wide where the dip is gentle. Sometimes the rock surface is nearly bare; elsewhere it is heavily forested with jack pines.

The Limestone Plateau

The crest of the Minnekahta ridge is the outer limit of a belt of more gently dipping Paleozoic limestones that cap the central Black

Hills uplift. In the east-central part of the dome, the Paleozoic sediments have been eroded, exposing the ancient gray schists and granites of the Precambrian basement over an area roughly 60 miles from north to south and 25 miles wide. This window is within an impressive escarpment of Paleozoic rocks.

Typically, the reddish brown Deadwood formation is at the base of the cliffs, and the pale gray Madison limestone caps them. Erosion of this window leaves a band of Paleozoic rocks only 6 to 8 miles wide on the eastern side of the Black Hills, as much as 30 miles wide on the western side. The wide portion of the plateau, cut by steep canyons walled with Madison limestone, is locally called the Limestone. The area is sparsely settled and contains some of the most spectacular scenery in the upper Midwest.

Many large springs exist along the foot of the east-facing escarpment at the eastern edge of the limestone plateau. They are generally at the base of the Mississippian limestone, which is underlain by the more dense Cambrian sandstones. The source of the water is rain and snow that falls on the eastern part of the limestone plateau. These springs are the headwaters of many of the major streams in the Black Hills.

The Central Black Hills

Precambrian rocks in the central Black Hills range in age from 1.7 to more than 2.5 billion years. They had eroded to a gently undulating plain by the beginning of Cambrian time. Erosion since the rise of the Black Hills has developed a rugged topography crowned by the mass of granite exposed in Harney Peak.

The temperature and pressure at which minerals in the Harney Peak granite crystallized suggest that the original molten magma crystallized at a depth of about 8 miles. Age dates place the time of crystallization at 1.7 billion years ago. Erosion of the overlying rocks exposed the batholith sometime before the sea flooded this area during Cambrian time, more than half a billion years ago.

The rising magma that was to become the Harney Peak granite sent out countless fingers of molten material, probing and injecting existing cracks and planes of weakness. Gradually, by lateral and vertical shoving, small individual blobs of magma worked their way up from deep within the earth's crust. Hundreds of separate bodies now make up the main granitic mass. Countless granite dikes fill old fractures within the surrounding schists. An oblique photo of Harney Peak from the air, or a drive along the Needles Highway, shows that the center of the batholith is a series of massive layers of granite, many with remnants of schist between them. These layers tend to steepen in dip in the outer part, producing a crudely domal structure.

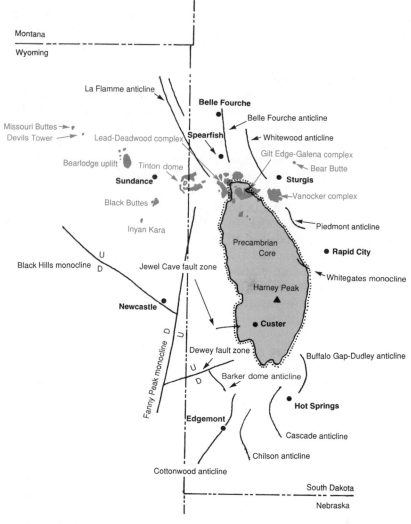

Axes of principal structures formed during rise of the Black Hills. U = *up side of fault,* D = *down side of fault.*

Pegmatites

Some of the Harney Peak granite is pegmatitic—that is, it is very coarsely grained with individual mineral crystals larger than 1 inch in diameter. This coarse grain size reaches an extreme in many smaller masses of rock of granitic composition that injected the metamorphic rocks around the granite. Some of these small pegmatites contain individual crystals tens of feet across.

222

More than 20,000 individual bodies of pegmatite exist in the metamorphic rocks surrounding the Harney Peak granite. Most are small, only a few feet thick and a few hundreds of feet long. The larger ones underlie local peaks such as Buckhorn and Thunderhead Mountains.

Pegmatites generally consist of minerals that contain more of many rare elements, such as lithium, tantalum, boron, and beryllium, than the main granite mass. Nevertheless, the general chemical and mineral compositions suggest that the main mass of granite magma was also the source of the pegmatite magmas. The pegmatite magma was rich in watery solutions, which lowered its melting temperature. So the pegmatite magma was the last molten rock to crystallize. It was enriched in all the elements that had no proper place in the minerals of the main granite. These included the rare elements.

Its large water content makes pegmatite magma extremely fluid, and that permits mineral grains to grow rapidly and very large. Tapered crystals in many pegmatites show that crystallization began at the contact with the adjacent rock, and proceeded inward toward the center of the pegmatite.

Black Hills pegmatites have been mined since the late 1800s. They provide mica, feldspar, and other minerals that are sources of rare elements such as beryllium, lithium, and tantalum. In addition to their commercial importance, the pegmatites have been a bonanza for mineralogists and amateur collectors. Sale of specimens is a thriving facet of the tourist industry in the Black Hills.

Pegmatite mines are a mineral collector's paradise. Black Hills pegmatites contain roughly 175 different mineral species. Most commercial mines contain 15 to 20 species, but some pegmatites contain many secondary minerals, derived from the original minerals. The Tip Top Mine east of Keystone contains more than 110 minerals, though most are in minute crystals best identified in the laboratory. An hour spent on the dumps of two or three inactive mines in the Keystone or Custer area will produce a collection that may include many of the following:

Quartz comes in many varieties. Clear and milky white quartz are very common in mineralized veins. Rose and smoky quartz are much less common.

Feldspar also comes in several varieties that differ in chemical composition. Most of the feldspar is in blocky crystals: white, pink, or red. The rare lithium minerals spodumene and amblygonite are white minerals that locally occur in commercial quantities. Early miners found a famous spodumene crystal 57 feet long in the Etta Mine at Keystone.

Mica can be spit into thin sheets. Muscovite mica is clear and very common; biotite mica is black. Some of the pegmatites contain the rare lithium mica lepidolite, which has a lavender color.

Tourmaline—mostly the ordinary black variety, schorl—is common in pegmatites. Some pegmatites contain a clear green variety of tourmaline with mica. Some of the pegmatites that are rich in lithium contain beautifully clear pink tourmaline.

Beryl is generally white, pale green, or brown, in hexagonal crystals. Individual crystals weighing many tons have been mined. It is mined as the main source of beryllium. The gem mineral aquamarine is blue beryl.

Apatite, a calcium phosphate mineral, occurs in a few pegmatite mines as small pale to deep blue masses. Other phosphate minerals include triphylite and lithiophyllite. Triphylite, when freshly broken, looks like a gray feldspar, but it darkens with a few hours of exposure to sunlight. Lithiophyllite is similar, but weathers to a distinctive purplish brown.

Columbite and tantalite are very heavy black minerals that occur in shiny black blades. Cassiterite is a tin oxide mineral that generally occurs in small grains.

Gold and Silver Mining

In 1874 the Custer military expedition to the Black Hills gave the first formal and official notice of placer gold, and the usual gold rush followed immediately. Those who came first staked claims along the streams; those who came later prospected for the mother lode, the bedrock source of the placer gold.

They soon found that the mother lode was lean, that most of the gold in it was present as impurities in other minerals, mostly the iron sulfides. The rock contained only minute blebs of free metallic gold. Millions of years of weathering and erosion of millions of tons of primary ore had freed the gold that the streams had concentrated in their placer deposits.

The miners learned that where ore veins had weathered they could recover a large percentage of the gold by grinding the ore and amalgamating the free gold with mercury. They called those weathered ore veins the free milling ores. They were brown, stained with rusty iron oxides that formed as the iron sulfide minerals oxidized, freeing the gold. The ores below the weathered zone were generally bluish gray or greenish; fine grinding and amalgamation would recover only 10 to 20 percent of their gold.

The miners and prospectors realized fairly early that two types of gold deposits exist in the Black Hills: the scattered ore bodies in the southern Black Hills and the huge ore body at Homestake, both of which were in rocks of Precambrian age. That mineralization happened nearly 2 billion years ago. The many smaller mines that operated in the northern Black Hills before World War I worked ore bodies next to vertical fractures in the much younger Deadwood formation. That gold was introduced during the period of igneous activity that accompanied the rise of the Black Hills 50 or so million years ago. Silver accompanied the gold in both types of deposit, more in the Tertiary deposits than in those of Precambrian time. Bullion consisting of a natural mix of gold and silver is called doré.

During the first 15 years of mining, nearly all the gold shipped from the Black Hills came from placer deposits or from the weathered ores. Some marginal ores were treated by roasting the crude ore with common salt, the chlorination process. Richer ores could be smelted, but that was expensive.

Simple grinding and amalgamation of oxidized ores can be done on almost any scale. Countless small companies built small mills on small properties. Most went broke. Consolidation of properties began on a large scale in the 1880s, leading to fewer and larger mills that could custom treat ore from many mines. Many miles of narrow-gauge railroads were built to connect mines and mills. The 25 years before 1910 were the heyday of mining in the Black Hills.

By the 1890s, the survivors included a few large companies and a scattering of small operators. Development of the cyanide process in the early 1900s made it possible to work the unoxidized primary ores, which were very low grade. Finely crushed ore is loaded into a huge vat and treated with a weak solution of sodium cyanide to dissolve the gold. The gold is recovered by treating the solution with zinc dust or by filtering it through charcoal. The cyanide is used again, so very little is lost to the environment.

About 30 cyanide mills were built between 1900 and 1905. Mines that could not supply enough ore to keep their mill busy, requiring some 300 to 500 tons of ore per day, could not operate at a profit. Few mines could produce that much ore, so the introduction of cyanide mills led to a new round of consolidations. Most of the new mills were idle by 1910, but those that succeeded paid regular dividends.

At that time, the federal government fixed the price of gold at $20.67 per ounce. Shortage of manpower during World War I and a rising cost of labor and materials after it ended closed most of the mines in the northern Black Hills. A few limped on to 1933, when the price of gold was raised to $35 an ounce. All gold mines were closed by government decree during World War II to force miners to shift to the vitally needed copper mines or to the shipyards. Very few mines were able to resume after the war. The great Homestake Mine was the major exception.

The Homestake Mine

The rush to Deadwood Gulch in 1875 nearly depopulated the southern Black Hills, where gold was first discovered near Custer. As soon as the placer ground along the streams of the northern Black Hills was staked, miners started prospecting for the bedrock mother lode. The first bedrock ore they found was in a hard ledge of conglomerate at the base of the Deadwood formation. It was a fossil beach placer, similar to the beach placer deposits miners found 20 years later at Nome, Alaska. The miners called this rock "cement ore" because it looked like concrete made with a stream gravel aggregate. They crushed the rock to free the gold, then recovered it by panning or by amalgamation with mercury. This discovery gave the miners a second type of placer deposit, but it still did not reveal the primary bedrock mother lode.

Moses and Fred Manuel and their partner Hank Harney discovered the Homestake lode on April 9, 1876, in a small outcrop of deeply weathered schist high on a hillside above what is now the big Open Cut. Crushing the rock yielded free flakes of metallic gold. This schist was the apparent source for the nearby placer gold, both in the ancient Deadwood formation and in the modern creeks.

The Manuel brothers worked their discovery on a small scale in the spring of 1876. In June 1877, they sold their 10 acres of claims to a California group that had made fortunes in silver in the Comstock lode at Virginia City, Nevada. The group incorporated the Homestake Mining Company in California on November 5, 1877. By July 1878, the company had installed an 80-stamp mill, and in January 1879 it offered Homestake shares on the New York Stock Exchange. It is by far the oldest continuously listed mining stock.

Although the outcrop of what was to become known as the Homestake formation was small, several companies started mining individual claims. Gradually, Homestake integrated its operations as it acquired those properties. Acquisition of mill sites and of water and timber rights rounded out the operation.

Miners soon recognized that the Homestake ore body trended southeast and that it plunged underground at 30 or more degrees from the horizontal. They did not fully understand the complexity of the ore body until after the company established a geological department in 1920. Further work finally established that the scattered pockets of ore all lay in one thin rock layer, sandwiched between two very thick rock units. All the rocks are of sedimentary origin, deposited in an ancient, shallow sea. Although now tremendously

The first mill for crushing gold ore, brought to the Black Hills by ox team in 1876. —Homestake Mining Co.

folded, sheared, and metamorphosed, the original sequence was as follows, with the Ellison formation being the youngest:

Ellison formation—Now a series of slates, mica schists, and quartzites, the sediments in this formation were originally shale, shaly siltstone, and sandstone, probably deposited on a submerged shelf. The original thickness was about 1,200 feet, though it has been thinned and stretched in many places.

Homestake formation—Now a series of iron-rich schists and slates, this formation was originally a chemically precipitated iron formation, largely iron carbonate and chert. The original thickness was between 60 and 90 feet, but later folding, shearing, and stretching have thickened and thinned it from zero to 400 feet.

Poorman formation—The Poorman formation, named for the good exposures in Poorman Gulch, consists of two very unequal units: The lower unit, now an amphibolite, was originally a mixture of

227

submarine lava flows and associated volcanic debris. The upper unit is a thin iron-rich schist high in carbonates, garnet, and graphite. Originally, it was a black shale rich in iron, sulfur, and carbon. The formation is about 5,000 feet thick in the mine area.

Repeated folding and shearing of the entire sequence left the outcrop of the thin Homestake formation resembling a piece of old-fashioned ribbon candy. Folding left the rocks tightly wrinkled, with the Homestake formation tremendously thinned on the flanks of the folds and probably thickened on the crests of the folds. Homestake geologists now believe that the gold rose into the rocks about 1.7 to 1.8 billion years ago when hot diluted water solutions rose along shear zones. Chemical reactions between the hot water and the rocks precipitated the gold in permeable zones within the Homestake formation. Earlier theories on the origin of the gold believed it was deposited from hot springs at the same time as the formation in which the gold occurs.

Homestake miners refer to the thickened and sharply folded zones in the crests and troughs of the elongated folds as ledges. The miners number the mineralized troughs, or synclines, with odd numbers, the barren crests, or anticlines, with even numbers. The mineralized ledges are, from east to west, the Caledonia, the Main, and odd numbered ledges 7 through 21. Only the Main and Caledonia ledges crop out at the surface. Discovery of the others happened as miners drilled and mined away from the original ledges.

Most of the ore is in elongate, rather tabular pods within nearly barren rock. The pods vary in size and make up a very small portion of the entire Homestake formation. Ore pods must be discovered and tested by diamond drilling far ahead of the mining operation. Most drilling is done from underground. Detailed geologic mapping in the underground workings and careful logging of drill cores makes it possible to make reasonable projections from one level to another.

The Homestake Mine is on a series of levels 150 feet apart. The lowest levels are more than 8,000 feet below the surface, more than a half mile below sea level. Two shafts, whose headframes are conspicuous on the skyline at the south edge of Lead, provide access to the mine. Each drops to a main haulageway at the 4,850-foot level. Several underground shafts, or winzes, drop from the haulageway level to the bottom of the mine. The two main shafts transport men, materials, and ore. One ventilation shaft takes in fresh surface air, another exhausts hot and humid air. Because rock temperatures in the bottom of the mine are hotter than 130 degrees Fahrenheit, large portable refrigeration units are needed to cool the air on the lower levels.

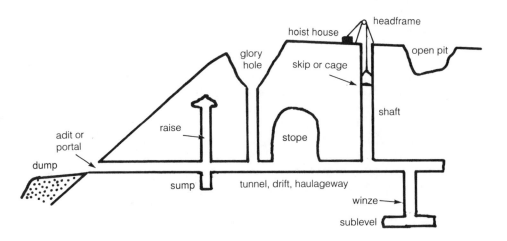

Hypothetical mine.

Miners open stopes as they mine the ore. Some stopes are small; others extend from one level to another. Mining techniques have varied over the years, and they differ between the individual ore bodies. Broken ore comes to the surface in chunks as much as 30 inches across. It is stored in huge bins to await milling.

The ore goes from the bins through a series of crushers that reduce it to fragments one-half inch or less in diameter. This crushed rock then goes to mills that grind it until 70 percent will pass through a 200-mesh screen. It emerges from the mills as a slurry of pulverized ore suspended in water. Machines called classifiers then separate the slurry into two fractions: the coarser fraction is "sand," the finer fraction is "slimes."

The sand fraction goes into a series of vats, each holding 750 to 780 tons, where a solution of sodium cyanide trickles through it. More than 90 percent of the gold can be dissolved by repeating the process several times. Zinc dust precipitates the gold from solution. Filtering removes the gold, which then goes to the refinery for final treatment.

The slimes also go into huge vats, where they are agitated in a solution of sodium cyanide to dissolve the gold. Activated carbon then absorbs the gold from the solution. The gold is then stripped from the carbon and sent to the refinery. The recovered gold from both processes is further purified, then cast into bricks that are 99.99 percent pure.

After the gold is leached out, the mills return the sand fraction underground to fill old stopes. They pump the slimes to a settling and storage area in Grizzly Gulch.

229

The last few decades brought a revolution in the cyanide recovery of gold and silver. Instead of grinding the ore to a fine powder and treating it in batches in leaching tanks, most gold mines now use a process called heap leaching. They crush but do not grind the ore, then spread it over an impervious pad. No building is necessary. Many small sprinkler heads spread a weak solution of sodium cyanide over the surface. The cyanide solution dissolves the gold as it filters down through the ore, then it trickles through large cylinders packed with activated charcoal made from coconut shells, which recovers the gold. After the gold is recovered, the charcoal is reactivated and reused. The cyanide solution is also reused.

Rich ores are still best processed by conventional milling, but the heap leaching process enables mining companies to recover gold from huge volumes of lean ores taken from open pits. Using this process, one company finds it profitable to process any rock containing over one sixty-sixth of an ounce of gold per ton.

The impermeable pads under the ore heaps are made with great care to protect the groundwater below and to prevent loss of gold. Most pads are several hundred feet across, and they slope toward a central collection point. Most are made of alternating layers of tight clay and plastic sheeting, commonly laid on an asphalt pavement. Some mines use pads once; others reuse them.

Cyanide leach pads, west of Lead, recover gold from crushed ore. The thin white lines above the trucks are hoses with spaced sprinklers to apply cyanide solution to the surface of the heap. —Jim Lessard, Wharf Resources

A single-use pad may be loaded with one or more layers of ore that reach about 10 feet thick and weigh perhaps a million tons. Leaching continues until most of the gold is recovered, typically a matter of some months. Then a second layer of ore is spread across the first, and the process repeated. The final pile may involve a succession of such layers. Finally, fresh water sprinkled over the heap removes the cyanide and other noxious compounds.

To reuse a pad, the company carefully removes the first bed of spent ore, then spreads a new layer. In either case, at the end of the mining and leaching operation, the mining company must reclaim the land according to a reclamation plan previously agreed upon between the company and the state. Bonding of each operation assures coverage of reclamation costs, should the company go broke.

Heap leaching started in the Black Hills in 1981, and several open-pit mines are in operation. None is readily visible from the major highways, but a trip to the top of Terry Peak will give you an excellent view of several of the operations. Nearly all of these mines are in the Deadwood formation and the nearby igneous intrusions. All involve properties first mined before World War I. The early prospectors found all of the gold districts.

Road Logs of the Black Hills

I-90
Rapid City–Wyoming Line
60 miles

Think of the Black Hills as a sort of geologic bull's-eye, with the oldest rocks at the center and successively younger formations making concentric rings around them. The more resistant formations stand as ridges that circle the Black Hills, the less resistant formations make valleys between those ridges.

Rapid City to Sturgis (28 miles).

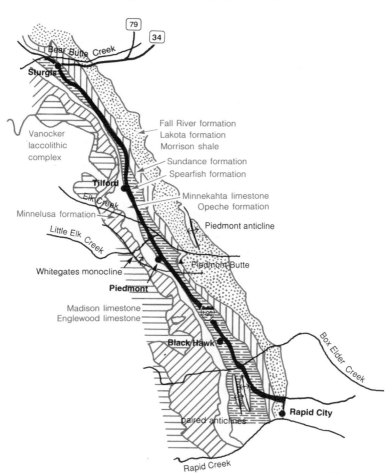

West

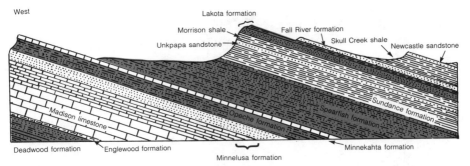

Lakota formation

Morrison shale

Fall River formation

Unkpapa sandstone

Skull Creek shale

Newcastle sandstone

Madison limestone

Opeche formation

Spearfish formation

Sundance formation

Deadwood formation

Englewood formation

Minnelusa formation

Minnekahta formation

East

Cross section through the Red Valley and Cretaceous hogback north of Rapid City near Exit 48.

The broad trough of the Red Valley follows the red, easily eroded Spearfish siltstone and the overlying soft sands and greenish gray shales of the Sundance formation all the way around the Black Hills. The high ridge or hogback on the outside of that circular moat is the younger but more resistant Lakota sandstone of Cretaceous age, which surrounds the Black Hills as though they were a medieval walled city. The inner ridge on the inside of the Red Valley is armored by the hard Minnekahta limestone. The major streams all cross the Red Valley at right angles. None follow it.

Between Rapid City and Sturgis

The Red Valley was eroded in stages. At several places along the route, remnants of high terraces with gravel caps may be seen They are remnants of old valley floors left as the streams eroded their beds to lower levels. Watch for those high terraces around Rapid City and Sturgis.

The interstate highway enters the Black Hills through the gap in the outer hogback on the north edge of Rapid City, then swings northwest and follows the Red Valley. The Chicago and Northwestern Railroad also used this easy route. Rapid City, Sturgis, Spearfish, and several smaller communities are all sheltered behind the hogback ridge.

Small yellow scars on the lower slopes of the steep inner side of the hogback between Rapid City and Sturgis mark quarry sites in soft Jurassic sandstones. The sand was used in the subgrade under the concrete surface of the interstate highway and as a filler in hot-mix asphalt.

Box Elder Creek is normally dry at this crossing, but during the flood of 1972 the stream flowed some 50,000 cubic feet of water per

Petrified logs on Piedmont Butte at Elk Creek Campground. —Jerry Teachout

second. The southbound span of the I-90 bridge soon choked with debris, creating a waterfall that undermined and destroyed the downstream northbound bridge.

Piedmont Butte is an isolated section of the hogback that stands between two stream valleys. The Lakota sandstone that caps the butte contains abundant petrified wood. Some geologists think it collected as logjams in stream channels in early Cretaceous time.

Dinosaurs

When O. C. Marsh, the famous dinosaur paleontologist of Yale University, was hunting fossils near the Black Hills in the summer of 1889, he noticed a large bone fragment on display in the post office of what is now Piedmont. He learned from the postmistress that it came from the north flank of Piedmont Butte. After visiting the spot and collecting further material, he recognized the fossil as a new genus of dinosaur that he later named barosaurus. A few years later,

234

View northwest across the Red Valley south of Piedmont. Ridge-forming Lakota sandstone is on the right. The tree-covered slope across the valley is Minnekahta limestone.

Professor George R. Weiland of Yale university was in the area collecting fossil cycads, a type of plant that was abundant during Jurassic time. He excavated the remaining barosaurus bones, but unfortunately the head was missing, as were some of the leg bones.

The Whitegates Monocline

Look on the skyline about 2 miles west of the interstate highway where it crosses Little Elk Creek for a sharp bend in the layers of Paleozoic rocks. That fold is the Whitegates monocline; some call it

A shear zone in deeper rocks probably formed the Whitegates monocline on Little Elk Creek during uplift of the Black Hills dome.

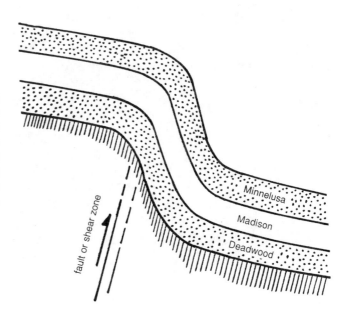

Red Gate. Deadwood sandstones dip east on the Whitegates monocline.
—J. E. Martin

*White Gate.
Madison
limestone dips
steeply east.*
—J. E. Martin

the White Gate monocline. You can most easily follow the fold by watching the white band of the Madison limestone, dipping eastward. Where it disappears into the canyon floor, the layers of rock are standing almost vertically. Early settlers named the narrows where the creek cuts through the white upper Minnelusa sandstone and the Madison limestone the White Gate. The deep reddish brown upper quartzite of the Deadwood formation forms the Red Gate a short distance upstream. Although the county road is washed out farther upstream, you can drive to within a few yards of White Gate from Exit 44. As in similar monoclines around the periphery of the Black Hills dome, the sedimentary rocks appear to be sharply flexed over a fault in the deeper basement rocks.

Pennsylvanian through Jurassic sedimentary rocks.

		Formation		Description	Thickness
		Morrison fm.		Green-to-lavender clay	0–220 ft.
		Unkpapa fm.		Fine varicolored sand	0–225 ft.
	Jurassic	Sundance formation		Redwater member: greenish gray shale with buff sandstone at top	
				Lak member: orange siltstone	
				Hulett member: yellow sandstone	
				Stockade Beaver member: greenish gray shale; fossiliferous	
				Canyon Springs member: yellow sandstone	300–450 ft.
	Triassic	Spearfish formation		Gypsum or anhydrite, white	
				Redbeds: sandy siltstone and clay, with gypsum	
				Redbeds with numerous thin layers of anhydrite or gypsum; called Goose Egg formation in Wyoming	300–700 ft.
	Permian	Minnekahta		Limestone, gray to pink	30–45 ft.
		Opeche fm.		Red siltstone and clay	75–135 ft.
		Minnelusa formation		Red-to-yellow sandstone	
				Thin dolomites	
				Yellow sandstone	
				Slabby dolomites, some sandy, gray to lavender	
				Red shale marker	
	Pennsylvanian			Massive dolomites and thin sandstones; some anhydrite underground; Leo sands may carry oil	
				Bell sand	385–800 ft.

Mud Flows and Boulder Fields

Extending fanlike into the Red Valley from the mouths of Stage-barn, Little Elk, and Big Elk Canyons are broad boulder fields, souvenirs of great floods on these drainages. Many of the boulders are as much as 3 feet in diameter, though a few may reach a length of 10 feet. Watch for them along the road between I-90 and the railroad tracks, where they were shoved out of the way during construction.

It seems likely that these big rocks came down the canyon in great mud flows that followed heavy rains. Mud is much denser than clear water, so it has correspondingly greater buoyancy, which can very nearly float rocks. A nice thick mud flow can carry boulders of astonishing size.

A detailed study of the Little Elk boulder field showed that most of the boulders came from Paleozoic rocks that were 2 or 3 miles above the mouth of the canyon. The degree of rounding that they attained in that short journey is phenomenal. A boulder count on the fan of Little Elk Creek from the creek bottom to just above the two gates showed 78 percent Deadwood sandstone, 11 percent Madison limestone, 6 percent Minnelusa sandstone or dolomite, and 5 percent Little Elk granite. The boulders are their natural color where exposed, but their undersides have a coat of white caliche, calcite deposited by groundwater.

The rounded boulders are arranged in crude ridges, and the intervals between the ridges mark separate distributary channels of nine or more successive floods. Weathering since the boulders reached their present position has accentuated the bedding in some of the

Boulder-strewn field in the Red Valley opposite the mouth of Little Elk Creek Canyon.

sandstones. Variation in depth of weathering of similar boulders in successive ridges and the extent to which lichens cover the rock show that the floods happened over a long time. No flood in historic times has approached the energy of these earlier floods. At a minimum, they were the kind of floods that come once in a century, or more probably, once in 500 or 1,000 years.

A short distance up Big Elk Creek is the old settlement of Calcite, where the Homestake Mining Company operated a lime plant as early as 1904. The raw material was originally Minnekahta limestone, but Homestake turned to other sources of lime in 1930 when the narrow-gauge railroad in the canyon was abandoned.

Vanocker Laccolith Complex

Between Tilford and Sturgis, the Red Valley narrows and the highway turns slightly east to go around a mountainous area formed by a cluster of large laccoliths. Sedimentary rocks dip steeply away from the individual intrusions. The igneous core of the largest laccolith northwest of Tilford is about 10,000 feet in diameter and more than 1,200 feet thick at the center. The core of the domes is a porphyritic rock similar to that of Bear Butte. An age date of 56 million years indicates that the intrusion occurred early in the rise of the Black Hills.

About a half mile north of Exit 34, the inner slope of the hogback is very unstable and marked by ancient landslides within the weak Jurassic shales. A highway cut made during construction work in 1963 upset the equilibrium of the slope, causing a large slide. Geologists and engineers of the highway department suggested that

Reconstructing the westbound lane of I-90, 3 miles southeast of Sturgis. The undulating topography on the skyline is an old landslide. —S.D. Department of Transportation

Gravel-capped terrace in the Red Valley southeast of Sturgis.

moving large volumes of shale from overloaded areas at the top of the slope to underloaded areas at the base could restore the equilibrium. The job was done, and no landslides have moved since.

Flat terraces in the Sturgis area mark an earlier level of excavation of the Red Valley. Bear Butte Creek imported the gravels that cap the terraces.

Between Sturgis and the Wyoming Line

Oyster Mountain, 2 miles northwest of Sturgis, is a high point on the Cretaceous hogback. It gets its name from slabs of rock on the flank that contain fossil oyster shells in the Sundance formation. Oysters are very common in Jurassic and Cretaceous sedimentary rocks; that must have been a splendid time for lovers of seafood.

Whitewood Creek, which crosses the highway just east of Whitewood, has recently been reclaimed after more than a century of pollution from mine wastes and sewage from Lead and Deadwood. A modern sewage treatment plant at Deadwood and diversion of mill tailings to a storage area south of Lead cleaned up the water. The stream once again supports aquatic life.

Whitewood Anticline

An anticline, sharply arched in the sedimentary rocks, widened the Red Valley at Whitewood. Fractures in the crest of the fold make the rocks more easily erodible. The fold, which formed when the Black Hills rose, continues northward in the sedimentary strata for about 30 miles. Close against the west flank of the anticline is a matching downfold, or syncline. It brings the Cretaceous hogback

240

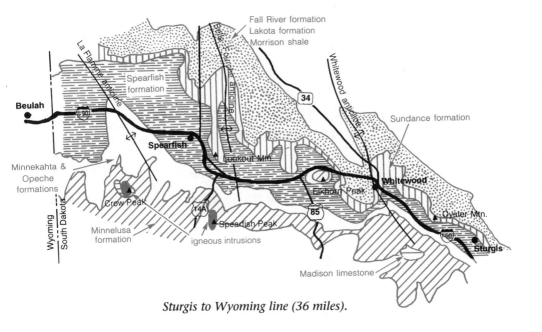

Sturgis to Wyoming line (36 miles).

south, so I-90 cuts through Jurassic and Cretaceous beds for about a mile before returning to the Red Valley.

Elkhorn Peak

The highway curves to follow around the base of Elkhorn Peak, a bald dome. The Minnelusa formation wraps over the crest of the dome, and steeply tilted triangular remnants of Minnekahta limestone stand among the trees around its base.

The dome of Elkhorn Peak is the roof over a blister of molten magma that invaded the base of the Deadwood sandstone. The igneous core is calculated to be more than a mile across and at least 800 feet thick.

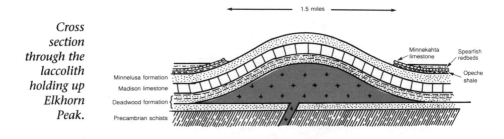

Cross section through the laccolith holding up Elkhorn Peak.

Spearfish Peak

The sedimentary rocks between Elkhorn Peak and Spearfish tilt gently down to the north, and that makes the Red Valley wider. Stream gravels conceal much of the red Spearfish formation. Ragged peaks that stud the skyline to the south belong to a cluster of igneous intrusions about 12 miles across. The rocks are mostly trachyte and phonolite, which contain much less silica than the rhyolite in the Vanocker complex to the east. Age dates range from 54 to 60 million years. Spearfish Peak is the high rounded mass near the west end of the group.

Lookout Peak

The sharp, isolated outlier of the hogback just east of Spearfish and north of I-90 is Lookout Peak. It contains very nice exposures of the Sundance formation, capped by a small remnant of Lakota sandstone. Near this peak in March 1887 Louis Thoen discovered a sandstone slab with the following messages crudely inscribed on its two flat surfaces:

> Come to these hills in 1833 seven of us De Lacompt Ezra Kind G. W. Wood T. Brown R. Kent Wm King Indian Crow all ded but me Ezra Kind. Killed by Inds beyond the high hill got our gold June 1834.

> Got all of the gold we could carry our pony all got by Indians I have lost my gun and nothing to eat and Indians hunting me.

Massive white gypsum at Spearfish lies on the Spearfish redbeds and is capped by yellow sandstone and shale of the Sundance formation. Lookout Peak is on the skyline.

Minnekahta limestone near crest of La Flamme anticline.

Spearfish is along Spearfish Creek, the major stream draining the northwestern section of the Black Hills. In 1893, the federal government established a fish hatchery at a large spring on the southeast side of town.

La Flamme Anticline

West of Spearfish, the interstate cuts through a broad upfold known as the La Flamme anticline. The upfolding brings the Minnekahta limestone to the surface. The outcrop shows gentle little folds and some small thrust faults. Look closely at the limestone to see its fine banding. Pairs of light and dark bands appear to represent an annual cycle of chemical precipitation of calcium carbonate from highly concentrated seawater. The formation contains no fossils except a few fish and some algal structures near the base that look something like cabbage heads.

Brilliant red cliffs of Spearfish formation north of I-90 near the Wyoming border.

McNenny Fish Hatchery

A mile or so north of the highway, and along a couple of miles on each side of the state line, a series of small, deep lakes dot the Red Valley. Artesian springs rising from the Minnelusa formation and perhaps the underlying Madison limestone have dissolved gypsum beds in the Spearfish formation and possibly in the Minnelusa formation. The overlying red shale collapsed into the solution cavities, opening sinkholes, some as large as several acres. The sinkholes are now spring-fed lakes. The Mammoth site at Hot Springs probably had a similar origin. The state of South Dakota operates a fish hatchery at one of these springs.

If you drive in to look at the lakes, notice in the fields along the access road, large angular blocks of quartzite resting on the redbeds. They are remnants of a silicified sandstone at the base of the Lakota formation, which by projection was about 900 feet above the present surface in this area. During the millions of years it has taken nature to erode 900 feet of shale and sandstone, these durable blocks have gradually descended to their present position with no appreciable wear.

Beulah, just across the state line in Wyoming, is along Sand Creek, one of the unspoiled streams draining the area along the state line. Ruins of an old waterpowered flour mill stand on the northwestern edge of town. Five miles south of Beulah, spring water issuing from the outcrop of the Madison limestone supplements the flow of Sand Creek. The recorded flow of the spring is 24 cubic feet per second, or 15.5 million gallons per day. The U.S. Fish and Wildlife Service once operated a fish genetics laboratory and hatchery just below the springs.

US 14A
Sturgis–Deadwood–Spearfish
46 miles

US 14A follows a loop from the Red Valley at Sturgis, through Deadwood, and back to the Red Valley just east of Spearfish. You can see the gently dipping Paleozoic rocks equally well whether you enter the loop at Sturgis or at Spearfish.

From the interstate highway, the road passes successively older sedimentary rocks, down to the Precambrian metamorphic rocks around Lead. Masses of igneous rocks that intruded the older rocks as the Black Hills rose about 50 to 60 million years ago greatly complicate the picture. Before the Black Hills rose, the rocks we now see

244

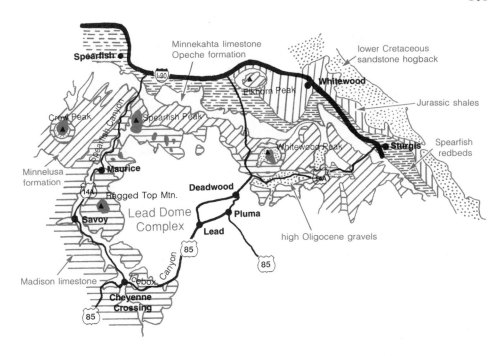

Sturgis to Deadwood to Spearfish (46 miles).

at the surface were buried beneath several thousand feet of sedimentary rocks, now mostly lost to erosion.

Between Sturgis and Deadwood

The route follows Boulder Canyon, which is very narrow at its mouth where the creek cuts through the pink Minnekahta limestone. Above the mouth of the canyon, nearly to the line between Lawrence and Meade Counties, the highway cuts through the yellow sandstones and gray-to-lavender dolomite beds of the Minnelusa formation. West of the county line, the canyon walls are gray Madison limestone, all crusty with lichens. Watch for the many small cavern openings in the limestone at and above creek level. It is as cavernous as Swiss cheese. Most of the year, the entire flow of the creek disappears into similar openings below stream level.

Boulders litter the floor of the Red Valley below the narrow notch through the Minnekahta limestone. During floods, Bear Butte Creek drops the boulders where it spreads out on the broad floor of the Red Valley.

Rainbow Arch

A strong anticlinal fold in the Minnelusa formation has a west limb that tilts down at an angle of about 70 degrees where you see it exposed at creek level. A short walk along the stream will take you across the entire Minnelusa formation, from the buried red soil on top of the underlying Madison limestone to the white sandstone at the top of the formation with its cross-bedding and elegant ripple marks. Here the formation is less than 500 feet thick. Watch for the west flank of the arch on the skyline when you drive toward Sturgis.

Older sedimentary rocks in the hogback around the Black Hills.

Age	Lithology	Formation, Member, Description	Thickness
Jurassic		Morrison: gray-green shale with dinosaur remains; some freshwater ostracodes	0–120 ft.
		Unkpapa: varicolored sandstone, nonfossiliferous	0–260 ft.
		Sundance: gray-green shale, massive yellow sandstones and thin limestones; represents two advances of western sea in upper Jurassic time; fossiliferous	375–450 ft.
Triassic		Spearfish: redbeds, consisting of silty shale and fine sandstone; red color due to iron oxide; numerous beds of white gypsum and anhydrite; nonfossiliferous	350–700 ft.
Permian		Minnekahta: thin-bedded pink limestone	30–55 ft.
		Opeche: redbeds similar to Spearfish	90–150 ft.
		Minnelusa: upper half is mostly fine-grained sandstone, with thin beds of limestone or dolomite; moderately fossiliferous	
Pennsyl-vanian		Lower half is mostly limestone or dolomite with thin beds of sandstone and red shale; moderately fossiliferous	400–700 ft.
Mississippian		Madison or Pahasapa: gray cliff-forming limestone and dolomite; cavernous; fossiliferous in places; mostly brachiopods, corals, and crinoids	450–600 ft.
Dev.		Englewood: pink shaly dolomite; fossiliferous	30–100 ft.
Ordo-vician		Whitewood: dolomite, buff, fossiliferous	0–70 ft.
		Winnipeg: roughlock silty dolomite member	0–30 ft.
		Icebox: green shale member	0–40 ft.
Cambrian		Deadwood: massive sandstone, limestone, and green shales; basal conglomerate carries fossil placer gold in places; forms red-brown cliffs and ledges	5–500 ft.

Slabby Minnekahta limestone overlying red Opeche shale. Beds dip westward into the Boulder Park syncline.

Boulder Park Basin

Just west of the monocline that forms the Rainbow Arch is a shallow structural basin, a syncline adjacent to an anticline. Its rim of Minnekahta limestone cradles a shallow remnant of red Spearfish shale, younger rock that supports few trees. Many people choose to live in this low park, mostly because it is beautiful, partly because they can get artesian water from the Minnelusa formation at a relatively shallow depth.

Just west of the Boulder Park basin, the highway passes a short road that cuts through red mudstone in the Opeche formation, which is sandwiched between the Minnelusa and Minnekahta formations. A very short side road leads to a quarry in the upper part of the Minnelusa formation. It once supplied molding sand to the Homestake foundry at Lead.

Mountain Meadow Surface

Several miles of roadway lying east of Deadwood, near the Oak Ridge Cemetery, cross a gravel deposit about 80 feet thick. It is a remnant of the Mountain Meadow surface, the highest of the Tertiary erosion levels. This accumulation of boulders and pebbles trends generally east, possibly preserving a segment of an early version of Whitewood Creek. That creek now flows north from Deadwood in a canyon 250 feet below the Mountain Meadow surface.

West of the Mountain Meadow gravels, the road drops into the present canyon of Whitewood Creek, passing good outcrops of the buff and red Whitewood dolomite. The black stuff at the bottom of

247

the hill, along Whitewood Creek, is smelter slag, a souvenir of the gold smelter that operated in lower Deadwood from 1891 to 1903.

Just east of the bridge across Whitewood Creek, north of the road, is a pyramidal rock outcrop that extends from the Ordovician Roughlock siltstone at the base up into the middle of the Madison limestone. It includes the best outcrop of the Englewood formation in the northern Black Hills. If you park east of the bridge, north of the road, you can walk around the base of the buff outcrops of the Roughlock and Whitewood formations. A trail, complete with poison ivy, climbs briefly, drops through a steep draw, then climbs onto a ledge of Whitewood dolomite. Above the ledge, which overhangs the creek, is a steep slope of gray shale that grades up into 30 feet of finely granular pink-to-lavender Englewood dolomite. The contact between the Englewood dolomite and the overlying gray Pahasapa or Madison limestone is usually identified by the color change from lavender to pale gray, but the color change shifts up and down several feet, even in this one outcrop. Farther up the hill is the remains of an early beehive lime kiln.

Along the south side of the highway at this same parking place, the sequence goes from the upper Winnipeg shale, through the Roughlock sequence, and up to the top of the Whitewood dolomite.

Between Deadwood and Cheyenne Crossing

The highway follows Deadwood Creek to a point north of Lead, where it abruptly turns southwest and ascends Poorman Gulch to the upper end of town. The contact between Cambrian and Precambrian rocks is well exposed on the gravel road along the southeast side of the rodeo ground. The road through town passes exposures of intricately folded schists, slates, and quartzites. They began their careers in Precambrian time as a very thick accumulation of shales, sandy shales, and sands, and then recrystallized into the metamorphic rocks you now see. The metamorphism happened during Precambrian time. Geologists divide the series into several unequal units, listed below with the oldest at the bottom:

FORMATION	THICKNESS	ROCKS
Grizzly formation	2,000 feet	schist and phyllite
Flag Rock formation	4,000–5,000 feet	basalt and chert
Northwestern formation	4,000 feet	mica schist and phyllite
Ellison formation	3,000–4,500 feet	quartzites and quartz-mica schist
Homestake formation	50–150 feet	iron-rich carbonate schist
Poorman formation	1,000–1,800 feet	slates and schists

Iron Dike

At the west edge of Deadwood, the road crosses a nearly vertical mineralized vein that trends north. The rusty appearance of its outcrop inspired people to call it the Iron Dike. In fact, it is a vein, not a dike. The mineralization is mostly iron pyrite, which was in great demand between 1891 and 1903 as a flux for the smelter at Deadwood. The pyrite carried one-tenth ounce of gold per ton, which was recovered during the smelting. The Olaf Seim Mine followed the vein north from the highway for about 800 feet. It is now a tourist attraction known as the Broken Boot Mine. The Montezuma and Whizzers pyrite mine drove a tunnel along the vein another 650 feet south of the highway.

Blacktail Gulch

The side road up Blacktail Gulch crosses a divide, then drops into the valley of Falsebottom Creek. Maitland was an important mining community from the early days until the last mine closed during World War II. It was not economic to reopen the mine after the war.

The mile and a quarter of Deadwood Creek between Blacktail and Poorman Gulches was a busy place in the late 1870s and early 1880s. Of five separate settlements strung along the creek, only Central City has survived. Many small mills were crushing cement ore, a fossil beach placer at the base of the Cambrian Deadwood formation. The name comes from its resemblance to concrete. Simple crushing freed the gold, which could then be concentrated by amalgamation using mercury. The cement ore mines, north of Lead, were high on the divide between Deadwood and Whitewood Creeks. The name Lead, pronounced "leed," refers to a stringer of ore that might lead into a larger and richer lode.

Cambrian-Precambrian Unconformity

A splendid exposure of the angular unconformity between the tilted Precambrian rocks and the flat Cambrian rocks that cover them is less than a half mile from the highway. Drive around the south side of the high school on Summit Street, and fork right on Pavilion Street. The outcrop is on the left, 500 yards ahead.

Very deeply weathered Precambrian rocks with steeply inclined beds lie beneath an old erosion surface planed right across their bedding. Horizontal layers of sandstone and conglomerate, the Deadwood formation, lie above the erosion surface. The age gap between the two rock sequences is more than 1 billion years.

The Precambrian rocks were deposited as sediments, then tightly folded as they were metamorphosed into slates and schists. Next, following more than a billion years of weathering and erosion, they were exposed at the surface. The buried erosion surface was the land

Flat-lying Deadwood quartzite resting on steeply dipping, deeply weathered Precambrian schist, on Houston Street in Lead.

surface at the beginning of Cambrian time, almost 600 million years ago, when the first animals with hard skeletons appeared on the earth.

Look at the left end of the outcrop for a rectangular adit cut into the basal quartzite. No doubt, the prospectors were looking for the gold that commonly occurs in the dolomite beds just above the quartzite. Hundreds of prospect holes in exactly the same geologic position dot the northern Black Hills. Look for the dike at the other end of the outcrop. Molten magma injected into a fault and crystallized there to form the dike. It displaced the sedimentary layers a few feet.

If you drive ahead a quarter mile to Mill Street, you can turn left and get an excellent overlook of the Homestake open-pit mine as you descend to Main Street in Lead.

After the junction of US 14A and SD 85, at the upper end of Lead, the highway passes the high school and drops through rusty red rocks of the Homestake formation, sandwiched between gray schists and quartzites of the Poorman and Ellison formations. Many pale rhyolite dikes cut the darker Precambrian rocks.

Near the turnoff to the Deer Mountain Ski Area, the roadcuts expose shales and sandstones of the Deadwood formation, which have been intruded at or near their base by Tertiary dikes and sills of pale igneous rock. The highway crosses a divide near the side road that leads north to Terry Peak. The lookout there offers a splendid view of the northern Black Hills.

Open cut of the Homestake Mine in Lead. Rhyolite dikes of Tertiary age cut steeply dipping Precambrian rocks. The dikes flatten at the contact with the Deadwood formation at the top.

West of the Terry Peak turnoff, the road follows Icebox Canyon and drops 1,000 feet in 3 miles to Cheyenne Crossing, where US 14A and US 85 diverge. In this interval, the road gradually descends through the Madison formation, through the buff-to-red Ordovician Whitewood dolomite, through green shales and siltstone of the Winnipeg formation, and partly through the thick Cambrian Deadwood formation. Most of the greenish gray shale outcrops are in the Deadwood formation.

A Mining Empire

West of the highway, and extending south for 2 miles from Nevada Gulch, is a block of patented mining properties originally assembled by the Golden Reward Consolidated Gold Mining and Milling Company between 1887 and 1918. By the end of World War I, the richest ore was exhausted, labor and other costs were increasing, and the price of gold remained constant, fixed by law, at $20.67 per ounce. The mine closed in 1918 after the company had produced about 1.3 million ounces of gold, a record second only to that of the Homestake Mining Company. When the mine closed, the property consisted of 440 mining claims, totaling about 3,963 acres, plus the mills and other facilities necessary to transport and process the ore. Mining resumed in the 1980s, after a great increase in the price of gold and development of cheaper mining and milling methods.

Between Cheyenne Crossing and Spearfish

The highway follows the canyon of Spearfish Creek. Between Cheyenne Crossing and Savoy, the lower canyon walls expose almost the full thickness of the Deadwood formation, which supports a good plant cover. The upper canyon walls are gray Madison limestone, covered with lichens. Between the two, the brown Whitewood dolomite typically makes a conspicuous ledge about 40 feet thick.

For about a mile, between Elmore and the mouth of Annie Creek, the highway cuts into a thick sill of igneous rock. The molten magma

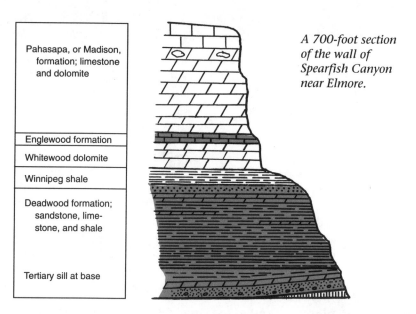

Pahasapa, or Madison, formation; limestone and dolomite

Englewood formation

Whitewood dolomite

Winnipeg shale

Deadwood formation; sandstone, limestone, and shale

Tertiary sill at base

A 700-foot section of the wall of Spearfish Canyon near Elmore.

intruded along or just above the unconformity between the Cambrian Deadwood formation and the Precambrian schists beneath it.

The stream in Rubicon Gulch could not cut through a sill of tough igneous rock within the Deadwood formation, so its water enters the main valley over Bridal Veil Falls. When a tributary stream cannot erode its valley fast enough to keep pace with the main stream, the result is a hanging valley that ends in a waterfall or cascade.

Between Savoy and the mouth of Spearfish Canyon, the sedimentary strata dip down to the north more steeply than the gradient of Spearfish Creek. One after the other, each of the formations exposed in the canyon walls disappears beneath the valley floor. About a mile below Bridal Veil Falls, the Deadwood formation dips below

Junction of Spearfish and Little Spearfish Canyons at Savoy. Slabby Deadwood formation is in the foreground. Aspens conceal the Icebox shale and Roughlock siltstone. Just above trees is the Whitewood dolomite, capped by the Englewood formation. The cliff is Pahasapa, or Madison, limestone. —J. E. Martin

Cliffs of Madison limestone in Spearfish Canyon. —S.D. Department of Tourism

Solution openings in Madison limestone, Spearfish Canyon. —S.D. Geological Survey

the valley floor, followed by the orange-and-brown Whitewood dolomite, then the gray Madison limestone, and finally the upper yellow sandstone of the Minnelusa formation. At the very mouth of
the canyon, the highway cuts through a narrow slot in the pink
Minnekahta limestone and emerges into the Red Valley.

Small Hydropower at Savoy

Little Spearfish Creek drops more than 1,200 feet between Savoy
and Spearfish. This makes it possible to pick up the water at the
upper creek level and redirect it through a gently dipping flume until
it reaches a point several hundred feet above the lower creek level.
Then the water drops through a penstock to a turbine at creek level,
where it generates electric power. Homestake Mining Company got
the necessary water rights in the early days and built such a system
just before World War I. It still operates during the months when
the stream has water. One generating plant operates at Maurice, where
the water drops 555 feet; another operates at Spearfish, where the
water drops 665 feet. You can see the flume high on the east canyon
wall between Savoy and Maurice.

Roughlock Falls in Little Spearfish Canyon, above Savoy. —S.D. Department of Tourism

US 16
Rapid City–Hill City
24 miles

At each end, this route gives a good view of colorful lower Mesozoic and Paleozoic sedimentary rocks. In the middle, between Rockerville and a point 8 miles west of Custer, you see a wide variety of metamorphic and igneous rocks.

South of Rapid City, the highway climbs 2 miles over the lower Cretaceous sandstones and clay of the Fall River and Lakota formations. The original road up this hill was both steep and tortuous, a popular place to test new cars during the first half of the century.

The layers of bedrock dip steeply down to the east, parallel to the slope—an arrangement perfect for landslides. When the road was rerouted through deep cuts, the layers of rock did indeed slide. One

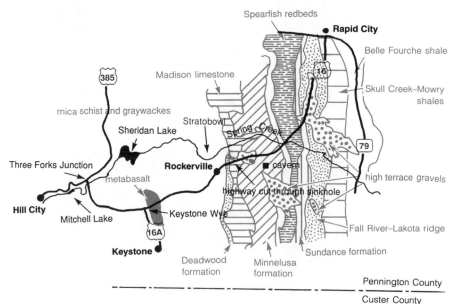

Rapid City to Three Forks Junction (24 miles).

slide dumped an estimated 26,600 cubic yards of Fall River sandstone and buff Fuson shale into one cut. Movement ceased after the uphill slope was regraded and proper surface drainage restored.

Old Erosion Surface

For several miles south of Rapid City, the highway follows the crest of the Cretaceous hogback along a remnant of an old erosion surface that beveled the Cretaceous and upper Jurassic sandstones, probably during late Eocene or early Oligocene time. The entire surface carries a thin blanket of stream gravel. The state geologic map shows the gravel as Oligocene, but that seems unlikely because this looks like a remnant of the old High Plains surface, which carries a cover of gravel deposited during late Miocene and Pliocene time. Furthermore, an ankle bone of a Pliocene camel was found in a shallow gravel pit west of the road.

View to the East

Late in the afternoon on a clear day, you can see the Big Badlands on the eastern skyline. Most of the high area is east of the Cheyenne River. West of the river, the Railroad Buttes appear as a long sinuous ridge capped by badlands clays and resting on Pierre shale. Here, in 1896, a party of scientists from Yale University unearthed the skele-

256

Stream-deposited clay and gravel capping the highest erosion surface just south of Rapid City.

ton of the largest Cretaceous sea turtle known, archelon, nearly 12 feet long. The skeleton was mounted and displayed at Yale University. Nearby were fragments of a still larger specimen, which had a skull 40 inches long.

Asymmetrical Thomson's Butte rises in the middle distance, 14 miles east of US 16. Twenty feet of coarse gravel cap 280 feet of Chadron and Brule clays, part of the White River group. They rest on Pierre shale.

Red Valley

Red sandstone and mudstone of the Spearfish formation are exposed in the rather steep eastern side of Red Valley. They were deposited during Triassic time, when red sediments accumulated on land in many parts of the world. Red and yellow Jurassic sandstones cap the redbeds. The wide valley floor is bright red shale, also part of the Spearfish formation, along with a few layers of white gypsum. The much gentler western slope is on dipping layers of Minnekahta limestone, with a bit of the red Opeche shale beneath.

Spring Creek, which flows across the Red Valley, loses its surface flow to sinkholes in the Madison limestone about 2 miles above the highway bridge. From this point downstream for several miles, the creek is normally dry except during the spring runoff and after intense summer thunderstorms. Where does the water go?

Water that soaks into the Madison limestone just upstream from the Red Valley moves downhill, toward central and eastern South Dakota. The overlying layers of red shale in the Minnelusa formation seal it into the Madison limestone, so it cannot leak to the surface. When a community in central South Dakota drills a well into the Madison limestone, it encounters artesian water that is under pressure because the aquifer is filling in the high country of the Black Hills.

You can see an excellent outcrop of that Permian shale in the red shale and siltstone of the Opeche formation north of the sharp curve where the highway starts to climb out of the Red Valley. The Opeche formation's position above the Minnelusa sandstone also explains the artesian water pressures in wells drilled to the Minnelusa formation around the flanks of the Black Hills. The maximum thickness of the Opeche shale is about 100 feet, but not all of it is exposed in this outcrop.

A period of weathering and erosion followed deposition of the Opeche formation before the sea returned to deposit the Minnekahta limestone. Weathering leached the upper few feet of the normally brick-red Opeche formation to shades of maroon and lavender. Many streaks of nodular limestone appear in relief on the lower part of the outcrop. These are a buried ancient desert soil, a zone of calcareous caliche similar to those so commonly seen in the arid southwest. Evaporation of soil water precipitates calcite that cements the soil into limy nodules.

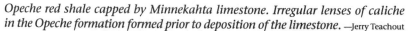

Opeche red shale capped by Minnekahta limestone. Irregular lenses of caliche in the Opeche formation formed prior to deposition of the limestone. —Jerry Teachout

Origin of the Black Hills Caves

By the end of middle Mississippian time, hundreds of feet of lime muds had accumulated on the sea bottom throughout the Rocky Mountain area. As the thickness and weight of the sediments increased, water was squeezed out and the muds compacted and recrystallized into limestone. Then the land rose, exposing a vast limestone plain. In this area, at least, the water table probably remained within 200 feet of the surface. Weathering and erosion began as soon as the new limestone surface was exposed. Rainwater, slightly acidified by carbon dioxide from the air, started dissolving the limestone. Much of the water entered fractures and moved down to the water table, enlarging the passages through which it passed.

When the water reached the water table, still slightly acidic, it either moved down, displacing the water already there, or moved laterally. This established groundwater circulation, with the water moving through vertical fractures and along horizontal bedding surfaces.

As the moving acidic water dissolved the limestone, it opened caverns. Periodic changes in the level of the groundwater table caused more active solution at successive levels, thus adding a third dimension to the cavern network. As time went on, a red clay soil, a residue of insoluble impurities within the limestone, began to accumulate on the limestone surface and on the cavern floors. From time to time, cave roofs collapsed.

Most of the collapsing caves eventually broke through to the surface, opening sinkholes through which water could enter, and into which red mud could filter. All drainage eventually shifted to the underground passages, and surface streams disappeared. A terrain of this sort is often called a karst topography, for a region in Eastern Europe. This process of solution and collapse is now going on in Florida, where newspapers frequently comment on holes suddenly opening at the surface, sometimes engulfing a car or a building.

When the land surface again sank during Pennsylvanian time, the inland sea again flooded the region. The incoming sea washed the red soil off the high spots, filled in the low areas, including the sinkholes, and buried the old karst landscape under layers of younger rocks.

Sometime between the end of the karst phase and the rise and erosion of the Black Hills, water chemistry changed, deposition succeeded solution, and calcite crystals started to grow on the walls of the old caves. These crystals grew with their long axis perpendicular to the walls. Thus, when we look directly at a crystal-lined face, we see the hexagonal outlines of the calcite crystals. The ends of the crystals may show three crystal faces meeting at a low angle, called a nailhead spar, or six faces meeting in a sharp point, called a dog-tooth spar.

Millions of years later, when the Black Hills rose and eroded deeply enough to expose the Madison limestone, water could again enter the cave system, further enlarging the caves. The higher elevation of the area encouraged the development of the artesian circulation that is still active. At many places in the old caves, you can see that the long crystals have dissolved on one side of a projection, but remain sharp on the other. Presumably, they dissolved on the side facing the flow.

The Black Hills stood high and the water table was low enough so that many of the caves drained. Water seeping out of the cavern walls then formed a baroque variety of cave ornaments: flowstone, dripstone, elephant ears, stalactites, stalagmites, popcorn, and so forth. Now that we are in an interglacial period in which the climate is much drier than it was during the ice ages, very little surface water seeps into the caves, and deposition of dripstone has virtually ceased.

Many rock falls, some huge, occurred when drainage removed buoyant support from the cave roofs. Segments of crystal lining also fell. Water dripping from the ceilings has partly cemented the fallen rubble into a solid mass. Now, you rarely see evidence of recent falls.

Caves breathe with changes in atmospheric pressure. When the outside air pressure drops, they exhale, discharging air until the pressure within the cave is the same as that outside. If the outside air pressure rises, caves inhale. Many cave openings have been discovered by the sound of air moving in or out through small openings or by a plume of condensing water vapor exhaled on a cold day.

It is possible to determine the approximate volume of a cavern system by measuring the volume of air passing in and out. Studies in Jewel Cave by Herb and Jan Conn and others have clocked wind speeds as high as 35 miles per hour through some constricted openings. Their rough calculations show that a 1 percent change in outside atmospheric pressure moves millions of cubic feet of air into or out of the cave.

Sitting Bull Cavern

The entrance to the cave is at the eastern edge of another segment of the old erosion surface, where a veneer of gravel overlies the truncated layers of Minnelusa formation. An anticline that trends north arches the Madison limestone to the surface in Rockerville Gulch, a mile south of the highway. Erosion of the gulch drained the cavern system in the upper part of the Madison limestone. The cave has an exceptionally thick lining of calcite crystals.

Condensation moisture on calcite wall crystals at Jewel Cave. —A. N. Palmer

Fossil Sinkhole

One mile northeast of Rockerville, the highway passes through a deep roadcut in the Madison limestone. On the south wall of the cut, the bedding planes in the gently dipping strata are continuous. The opposite wall is a section through a collapsed cavern filled with blocks of limestone and old red soil. The jumble is partly recemented with calcite.

The sinkhole existed before the Minnelusa formation was deposited, about 320 million years ago. After the Black Hills rose and eroded down to the Oligocene surface, clay washed into the exhumed cavern. Screening of some of the later clay filling has yielded rodent teeth of late Oligocene age. They help date the erosion surface.

Stratobowl

The Stratobowl was formed by a deeply entrenched meander of Spring Creek. The bottom of the bowl, covering several acres, lies 500 feet below the upland surface. A vertical fault that places Madison limestone against the lower part of the Deadwood sandstone cuts across the bowl, but its trace is not readily visible. It vertically moved the rock about 300 feet.

Roadcut through a filled sinkhole in the Madison limestone near Rockerville. Beds in the west edge of the sinkhole contrast with broken limestone debris in the sinkhole.

In 1934 the National Geographic Society and the U.S. Army Air Corps jointly sponsored a manned balloon flight, which they planned to launch from the Stratobowl. They canceled the original flight when the balloon split during inflation. The following year, at dawn on a cold November day, they launched another balloon, which measured 325 feet from gondola to top. During an eight-hour flight of 250 miles, the helium-filled balloon rose to a record height of 72,395 feet. It landed a few miles south of White Lake, South Dakota. This altitude record for manned flight stood until the age of manned space vehicle travel.

Rockerville

Rockerville was the site of a placer mining boom in 1876. Flakes of gold, found in high-level gravels and in the soil, apparently came from weathering and erosion of a fossil beach placer at the base of the Cambrian Deadwood sandstone.

As usual, the first mining was done with gold pans. The miners quickly replaced these with more efficient rockers. A rocker is a box on rockers, a bit like a baby cradle. It was standard placer mining equipment in those days, especially in districts where water was in short supply. Miners shoveled gravel into the box, added water, then rocked it. The sloshing concentrated the heavy gold in the bottom of the box. Traces of an old mule trail show that, before construction of the Rockerville flume, the miners packed water from the creek for their rockers.

Gold rocker used by Gold Bug Nelson, an old-time prospector and miner, on Slate Creek during the depression years of the 1930s.

Rockerville Flume

A large wooden and earthen flume, built in 1880, brought water to the dry gravels. To get the elevation to carry water out of Spring Creek Valley and into the upper part of Rockerville Gulch, the miners placed their diversion dam just above the present inlet to Sheridan Lake. Although the air-line distance is only 6 miles, the flume followed the contours of the land for almost 18 miles. Segments of the route are now hiking trails.

Flume leakage, financial mismanagement, and other problems beset the project. In 1880, Rockerville had 300 permanent residents and even more transients. Within a few years, the flume was in disrepair, and so was Rockerville, shriveling below the notice of the 1890 census. Rockerville Gulch produced about 20,000 ounces of gold, not much considering all the labor and investment that production entailed.

Keystone Wye

Take US 16A if you wish to go to Keystone and Mount Rushmore. Just before the highway reaches the Wye, watch for outcrops of dark green metamorphosed basalt. The three-level overpass is built of laminated timbers, rather than steel or concrete.

Along the next few miles, the highway skirts the north side of the Harney Peak batholith. Rocks are metamorphosed muddy sandstone, chert, and basalt. In the metamorphic rock between the road and the granite, thin veins of quartz and mica carrying small amounts of tungsten minerals cut across the sedimentary rocks. They produced a considerable amount of tungsten ore during World War I, a negligible amount during World War II. Only large government subsidies during national emergencies made mining profitable.

Three Forks

Several early gold mines worked steeply inclined quartz veins within the schists near the junction of US 16 and US 385. Some of the gold was in flakes of native metal, more existed as an impurity in pyrite and arsenopyrite, which contains arsenic as well as iron and is almost white. Those sulfides were oxidized near the surface, so their gold could be recovered by grinding and amalgamation with mercury. But the fresh ores at depth required smelting. Some of the better known mines were the J.R., Golden Slipper, Forest City, and Joe Dollar.

Mitchell Lake

Between Three Forks and Hill City, a dam across Spring Creek impounds a small lake next to the road. Opposite the lake, the road swings around an outcrop of graywacke. New cuts reveal a beautiful sharp fold in the bedded sandstone, lying almost flat like a folded omelet. A setback of the cut, to catch falling rocks, lies exactly on

Fold in Precambrian rocks in roadcut above Mitchell Lake about 1.25 miles east of the forest service ranger station at Hill City. Elongate blobs were originally limestone concretions. Steeply inclined white lines are drill holes.

White pegmatite from the Harney Peak granite intruded into mica schist.

the horizontal axial plane of the fold. Graded bedding shows that the older rocks are in the core of the fold, so it is an anticline.

Tin and Tungsten

Hill City is in a broad valley underlain by weak mica schist full of red garnets. Its fortunes vary with the demand for forest products and minerals. In 1883, miners identified tin and tungsten minerals in quartz veins and pegmatite dikes in the area around Keystone and Hill City. The metals evidently arrived with the Harney Peak granite.

With financial support from British interests, the Harney Peak Tin Mining, Milling, and Manufacturing Company was incorporated in 1884. In the next few years, miners staked, purchased, or leased more than 2,000 tin claims, and many were patented. Companies built one tin mill at Keystone and another at Hill City. The ore mineral was the tin oxide, cassiterite. The company failed in 1894, probably because it had grossly overestimated the amount of available tin ore. Many of the homes in the Hill City area are on mineral claims originally owned and patented by the Harney Peak Company.

No real interest in tungsten developed until imports were cut off in World War I. The principal ore minerals were the iron and manganese tungstates ferberite, wolframite, and hubnerite. A little interest in tungsten properties revived during World War II, but that brought no appreciable production.

US 16
Hill City–Custer
14 miles

Between Hill City and Custer, the route skirts the western side of the Harney Peak Range. Although the road is almost entirely on schists and other metamorphosed sedimentary rocks, massive intrusions of pegmatitic granite are the most conspicuous rocks, especially east of the highway. Some gold occurs in quartz veins and in shear zones within the metamorphic rocks. A few small prospects developed into small mines, mostly between 1900 and 1940. Within a short distance of the highway are remains of the diggings and dumps of the Clara Belle, St. Elmo, Minnie May, Old Bill, and Grand Junction Mines.

Crazy Horse Memorial
In 1947 Korczak Ziolkowski conceived the colossal equestrian statue of Crazy Horse, the great chief of the Lakota Sioux. He started drilling and blasting the pegmatitic granite in June 1948. Korczak died in 1982, but his family carries on the work. The rock and the scale of the monument are like those of Mount Rushmore. At the present rate of progress, its completion will require many years.

Custer Mica
The town of Custer began to grow immediately after gold was discovered in French Creek, before the area was open to white settle-

Calamity Peak, a granite knob east of Custer, is named for Calamity Jane.

ment. Gold miners opened the Crown pegmatite mine in 1879 to produce sheet mica for window glass, lamp and lantern chimneys, and stove fronts. Sheet mica of varying size and quality occurs in the outer zones of some of the more complex pegmatite bodies. Normally, it is not competitive with imported mica from India, Madagascar, or Brazil.

Nevertheless, the Custer area produced a limited amount of excellent sheet mica during wartime, when prices were highly subsidized. During World War II, the government brought in skilled technicians from India to train the local craftsmen to split, trim, and grade mica to industry standards. Under peacetime conditions, miners save scrap mica and beryl, which is incidental to mining feldspar. Mills grind the mica for use as filler in tires and other rubber products, as coating for roll roofing, as lost circulation material for oil well drilling, and even as phony snow for Christmas decorations.

US 16
Custer–Wyoming Line
27 miles

This short segment on the southwestern flank of the Black Hills uplift is an instructive drive. The absence of dense stands of timber makes outcrops obvious, and the rough terrain has required many roadcuts into the bedrock. Fortunately for geologists, the highway department has not regraded all of the roadcuts, covered them with soil, and planted them to grass. This route has three segments: an eastern leg in Precambrian sedimentary rocks, a central section in Paleozoic limestones, and a western leg in the open terrain of the Red Valley.

Precambrian Sedimentary Rocks

In Custer, the highway crosses the major Grand Junction fault, but it takes careful use of geologic maps to locate its trace. Rocks east of the fault are micaceous schists. On the west, the road crosses metamorphosed sandstone, or quartzite, for several miles. It finally disappears beneath a cover of Paleozoic limestones that still lie almost perfectly flat.

Mounds of earth and rocks mark old placer mines in the broad valley of French Creek south of the highway. Some of the disturbances date back to the gold discovery days of the 1870s, but the more obvious workings are the result of dredging during the 1930s.

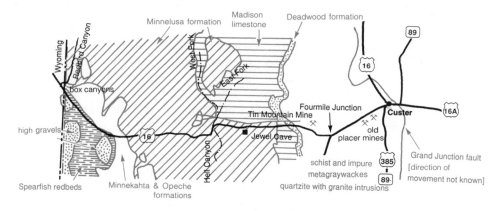

Custer to Wyoming line (27 miles).

Watch north of the road for the eastward increase in the number and size of granite outcrops. That reflects the approach to the main mass of Harney Peak granite just north of Custer.

The Tin Mountain Mine worked a complex pegmatite body that has produced commercial quantities of several lithium minerals and the rare cesium mineral pollucite. To see it, take Forest Service Road 287 north from a point 5.5 miles east of Jewel Cave or 3 miles west of Fourmile Junction. Go in about a quarter of a mile, then turn right for half a mile on an unimproved trail.

Paleozoic Limestones

Just west of Tin Mountain Mine, US 16 crosses the whole Paleozoic sedimentary rock section, from the Cambrian Deadwood formation at the eastern end to the Pennsylvanian and Permian Minnelusa formation at the western end.

The Cambrian Deadwood formation is a dark red sandstone, poorly exposed in a fault slice just west of the valley of Lightning Creek. It was found to be about 180 feet thick in a water well at Jewel Cave National Monument. Ordovician, Silurian, and Devonian rocks are missing in the southern Black Hills, presumably because they were eroded before the inland sea again flooded across the area in Mississippian time. The overlying Madison limestone, between 450 and 500 feet thick, forms massive light gray cliffs. Roadcuts expose thinly bedded sandstones and dolomites of the Minnelusa formation, which is about 625 feet thick.

About 1900 someone discovered Jewel Cave, which is in the upper half of the Madison limestone. The original opening was in the east wall of Hell Canyon about 75 feet above the creek. The present public entrance is about a mile east of the original opening, through

268

a shaft started in the lower part of the Minnelusa formation. Cave levels are at depths of 200 and 265 feet. The cave and a large surrounding area were set aside as Jewel Cave National Monument in 1908.

Detailed mapping of Jewel Cave has been in progress for many years. A glance at the accompanying map shows that solution of the limestone to form the cave has been principally along parallel vertical fractures that trend about 58 degrees east of north. A less well-defined system of cross fractures offers connections between the main

Frostwork in Jewel Cave. Crystals are about 0.5 inch long.
—Steve Baldwin

Dogtooth spar crystals, which are calcite crystals, in Jewel Cave. —Steve Baldwin

passageways and gives the cavern a network pattern. More than 83 miles of passageways have been mapped within an area roughly 12,000 feet across from east to west and 5,300 feet across from north to south.

A zone of faults with an east-to-west trend continues for 7 miles. The raised side of each fault is on the north. This becomes obvious when you are driving in a roadcut of Minnelusa sandstone and dolomite and see cliffs of the underlying Madison limestone towering above the north side of the road. The vertical displacement is at least 120 feet; the fault moved while the Black Hills were rising. Look in the high cuts to see how the thin slices of rocks are distorted and tilted within the fault zone.

Map of the openings in the east end of Jewel Cave. —Jewel Cave National Monument

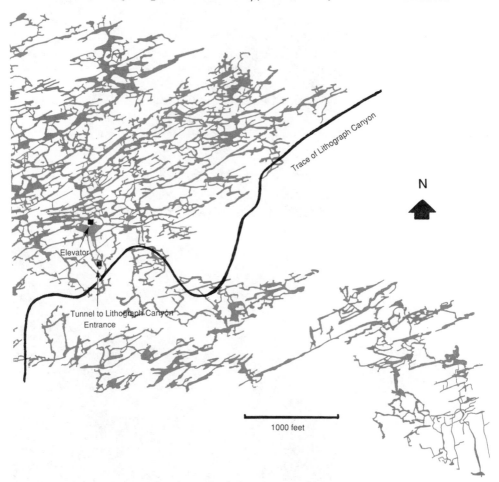

N

1000 feet

Terra rossa is a deep red clay soil that develops as limestone weathers in a warm and wet climate. Such a fossil soil developed on the Madison limestone during early Pennsylvanian time, before the area again sank below sea level. The incoming Pennsylvanian sea washed the clay off the higher parts of the eroded surface and into low spots and sinkholes, so the thickness now found is highly and unpredictably variable. Watch for conspicuous outcrops of red clay in roadcuts along both sides of Hell Canyon near the western boundary of Jewel Cave National Monument.

Small but beautiful fortification agates are in some of the chert masses in the lower Minnelusa dolomites in both Hell and Tepee Canyons. The very fine banding consists of alternating layers of red, black, and white silica. This occurrence gives a clue to the bedrock origin of the more famous Fairburn agates that are known only from gravels along French Creek east of Fairburn.

Red Valley

As you enter the Red Valley, the country opens, with long dip slopes of Minnekahta limestone to the northeast. Outcrops of red Spearfish shale are conspicuous south of the highway. Jurassic and Cretaceous sandstones form the high hogback on the skyline to the south and west. The redbeds are only about 350 feet thick on the southwestern flank of the Black Hills, but because the dip of the rocks is so gentle, the Red Valley is exceptionally wide.

If you walk up a box canyon, you will find that it suddenly ends in a steep slope, with rimrock closing around the top. Box canyons typically form where a thin capping of hard rock overlies an easily erodible formation. In the Black Hills, the most typical box canyons are cut into the Opeche red shale beneath a rim of Minnekahta limestone. Many excellent examples exist along the Red Valley just east of the state line.

Enter Redbird Canyon on Forest Service Road 376, which leaves US 16 a half mile west of the state line. Excellent outcrops of the Minnelusa and Madison formations permit detailed study. The Minnelusa formation reaches a thickness of more than 700 feet, but it further thickens underground to the southwest.

A monocline is a fold in which the rocks dip in one direction, rather than in opposite directions as in an anticline or syncline. The north-south trending Fanny Peak monocline, which dips steeply west nearly parallel to the border between South Dakota and Wyoming, is one of the choicer geologic structures in the Black Hills. The highway crosses it about 5 miles west of the state line, where the folding tilts the lower Cretaceous sandstones. North of the highway, the monocline gradually swings over to the South Dakota side of the line, where it conspicuously tilts the Minnelusa formation, forming a high red-and-yellow ridge on the skyline. To the south, the structure continues into the eastern edge of the Powder River basin.

US 16A
Keystone Wye–Iron Mountain
11 miles

This short segment of highway passes an amazing number of geologic points of interest in the Keystone area.

At the junction of US 16 and US 16A, schists heavily stained with rusty iron oxides appear in roadcuts just south and west of the Wye. They mark the mineralized shear zone that hosts some of the veins that the gold mines worked at Keystone. Principal minerals accompanying the gold ore were pyrite, pyrrhotite, and arsenopyrite. The tightly folded and metamorphosed Precambrian rocks around the Wye were originally impure sandstones and shales. Masses of dark green metamorphosed gabbro, muddy sandstone, shale, basalt, and chert tend to be associated with the iron formation.

Just south of the junction, schists with nearly vertical grain mark a mineralized shear zone that passes southward through gold mines and prospects around Keystone. Between the Wye and Keystone, the road follows Buckeye Gulch. The mostly ancient sedimentary rocks have metamorphosed into mica schists and quartzites. A variety of granite intrusions and some veins of white quartz cut them. Partway down the hill, the bypass around the highway tunnel is a

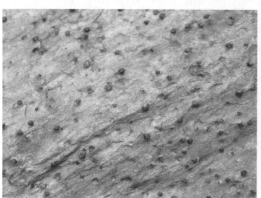

Garnet schist from Keystone. The garnets are up to about one-eighth inch across. —S.D. Museum of Geology

Staurolite and garnet schist from Keystone. —S.D. Museum of Geology

272

good place to examine the schists. Look for tiny red garnets and crossed crystals of staurolite in some of the schist layers. The parking lot next to the road at the foot of the hill is another good place to look.

At the traffic light, a turn west takes you along Battle Creek to Hill City. A turn east takes you down into Old Keystone and the remains of several gold mines. South from the light, US 16A follows Grizzly Bear Creek through New Keystone. At the south end of town, a road leads east from the heliport and chair lift to a group of pegmatite mines and their dumps. These are a focal point for mineral collectors and rock hounds. The road splits a mile south of the traffic light. US 16A continues to the crest of Iron Mountain; SD 244 leads to Mount Rushmore.

Keystone Gold Mines

Keystone was originally a mining town, and it has suffered the booms and busts of the business. In 1876, miners discovered placer gold in Battle Creek and gold in veins and shear zones within the schists. A few miners stayed to recover the placer gold, but the bedrock veins were small and hard to work, so Keystone was largely bypassed in favor of the rich diggings around Deadwood.

In the 1890s, some of the hard-rock miners gradually drifted back to Keystone to start developing the abandoned prospects. Four principal mines operated intermittently until about World War I: the Holy Terror, the Bullion, the Keystone, and the Columbia. Gradual consolidation of the mines and their connection by deep drifts permitted hoisting of all ore through one shaft. You can see the old mine dumps along the main street in Old Keystone.

Headframe, ore bins, and mill of the Holy Terror Mine in Old Keystone.

The Holy Terror is the best known of the mines; its abandoned shaft house and ore bins stand high on the hillside above the old town. According to local legend, an old prospector's wife was disgruntled because he spent most of his time hunting a bonanza, so he promised to name his future discovery after her. When he found a vein of white quartz glittering with native gold, he named it the Holy Terror. That mine followed the quartz vein down 1,200 feet, but heavy inflows of water and the litigation that followed a bad mine accident plagued the operation. It closed in 1902.

The Keystone ores contain a lot of arsenic. When the price of arsenic soared because of its use to control the boll weevil in the early 1920s, the company built an arsenic recovery plant in lower Keystone. Almost immediately, someone discovered another way to control the pest, and the price of arsenic collapsed.

Dewatering of the Holy Terror Mine was briefly successful about 1940, but further geological work did not reveal enough ore to justify reopening the mine. The flood of 1972 eroded part of the Holy Terror dump, and people later found specimens of native gold that had washed from it into Battle Creek.

Keystone Pegmatite Mines

Large masses of coarsely crystalline granite and pegmatite intruded the somber gray schists around Keystone. The granite was of no interest to the early gold miners, but as the gold boom subsided, the prospectors looked closer at some of the minor minerals enclosed within it.

Placer miners had identified the heavy tin mineral cassiterite in 1876. It was a nuisance that plugged their sluice boxes and was hard to separate from gold. The origin of the stream tin became clear in 1883, when small grains of cassiterite turned up in the granite that formed the outer zone of some of the pegmatites at Keystone. That started a tin boom.

The following year, columbite and tantalite, other heavy black minerals that plagued the placer miners, appeared in a deeper zone in the pegmatites. Mineralogists identified some of the lithium minerals at about the same time, but it was 1893 before they recognized spodumene from the Etta Mine as a commercial ore of lithium. They recognized beryl by its hexagonal crystal form, but commercial interest had to wait for a market to develop during World War II.

Commercial pegmatite mining in the Keystone area started in 1883. The level of activity varied widely with fluctuating market demands. The big boost in pegmatite mining came in 1923, when reduced freight rates permitted local feldspar to enter eastern markets. Reworking the old dumps is the only current activity. White

Large beryl crystals in coarse-grained granite from the Peerless pegmatite mine in Keystone. —D. M. Sheridan, U.S. Geological Survey

quartz is in demand for ornamental stone and for glass for fiber optics. Miners hand sort small quantities of cassiterite, columbite-tantalite, and beryl, and sell them locally. Many of the inactive mines are open, and collectors visit them regularly. The Dan Patch Mine is 2 miles west of the stoplight in Keystone; the Etta Mine is less than a mile east of the heliport and chair lift in the south end of town; the less accessible Bob Ingersoll Mine is 2 miles northwest of town.

Mount Rushmore

Mount Rushmore National Memorial lies along the northeastern edge of the Harney Peak granite batholith. The batholith is several miles across and extends to an unknown depth. Within the memorial boundary we see mica schist intruded by many granite offshoots from the batholith. The schist was originally mud and impure sand on an ancient sea bottom. Heat and pressure metamorphosed it into a slabby rock consisting mostly of mica and quartz. It splits parallel to the mica flakes, so the normally gray rock sparkles in reflected sunlight.

The granite in which the great heads were carved consists of small grains of quartz, white feldspar, and mica. It crystallized from molten magma at a depth of about 8 miles. As it cooled, it shrank somewhat and developed shrinkage cracks. Later, more molten material injected the cracks, forming granite and pegmatite dikes within the granite. These fracture fillings now show up as white streaks on the foreheads of Washington and Lincoln.

275

Fingers of white pegmatitic granite that invaded fractures in the schist at Mount Rushmore.

Faces of Mount Rushmore. —National Park Service

Deep erosion of the rocks that once covered Harney Peak exposed its granite long before the late Cambrian sea flooded the area and deposited the Deadwood sandstone about 550 million years ago. Throughout the southern Black Hills, the Deadwood sandstone lies on a gently undulating surface of weathered granite. The rise of the Black Hills roughly 50 million years ago caused erosion of about 7,500 feet of overlying sedimentary rocks. Continued weathering has left the very resistant granite standing as peaks, with the softer schists forming the valleys.

The great carvings of the presidents are in a thin sheet of granite several hundred feet thick, which cuts across the older schist. Old Baldy, 4,500 feet to the north, is a similar sheet of granite. We see the granite's somewhat irregular base as a color change below the bust of Washington, where homogeneous white granite overlies bedded gray schist. Below the schist is another rather uniform tongue of granite about 50 feet thick, then below it down to the top of the talus slope are more inclined beds of schist.

Iron Mountain

US 16A is strictly a scenic highway. In a horizontal distance of about 2 miles, it climbs from Keystone to the crest of Iron Mountain using many switchbacks, pigtail bridges, and tunnels. The main bedrock is mica schist invaded by countless masses of Harney Peak

Parallel intrusions of granite high on the flank of the Harney Peak batholith. —Roy Roadifer

granite. Tunnels along the route were carefully oriented to provide dramatic views of the great carvings on Mount Rushmore. From the top, you can see over much of the southern Black Hills and the adjacent prairies.

Iron Mountain gets its name from a large outcrop rich in iron that runs from the crest a short distance down its southern flank. The iron oxide minerals include an impure hematite called ironstone, as well as the hydrous iron oxide goethite. You can think of them as rust. Some of the goethite is beautifully iridescent. Layers of quartzite derived from chert and a thick mica schist sandwich the rock layers that contain the iron. The beds, now strongly folded and faulted, tilt at angles of 40 to 70 degrees.

The iron deposit was originally beds of iron carbonate and layers of graphitic schist rich in iron pyrite, which weathered to iron oxides. Some of that weathering happened before the Deadwood sandstone was deposited, some after the Black Hills rose and eroded to their present form. The base of the Cambrian beds, if restored over this area, would coincide with the top of Iron Mountain, so it must have been part of the old erosion surface on which the Deadwood sandstone was deposited.

Miners call this type of rusty outcrop a gossan. Gossans commonly form over sulfide ore bodies that contain other metals besides iron and so may mark the existence of valuable ore bodies below. Careful analysis of the Iron Mountain gossan fails to show significant traces of metals other than iron, so it probably does not mark buried riches. The old Cuyahoga Mine, farther down the slope, worked a bed of sulfide ores for pyrite and arsenic.

US 18
Maverick Junction–Wyoming Line
45 miles

The route enters the Black Hills through the Fall River Canyon water gap and follows the open Red Valley to the junction of US 18 and SD 89 at the former railroad settlement of Minnekahta. It climbs nearly 1,000 feet to get over the sandstone hogback to reach Edgemont, then stays on somber marine Cretaceous shales to the state line.

Between Maverick Junction and
the East Edge of Hot Springs

The highway winds through Fall River Canyon and traverses the geologic section from lower Cretaceous gray shales to the colorful Triassic redbeds at Hot Springs. Just above Hot Springs, two small spring runs, Hot Brook and Cold Brook, converge to form Fall River. The many warm springs along its course through the town supplement the flow. A mile below the mouth of Fall River Canyon, the stream enters the Cheyenne River, so the total length of the Fall River is less than 8 miles.

Maverick Junction to Wyoming line (45 miles).

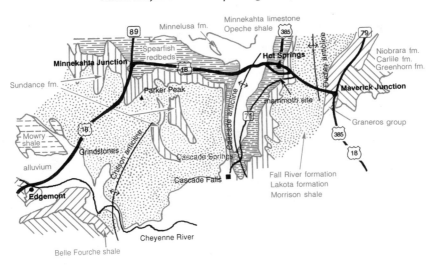

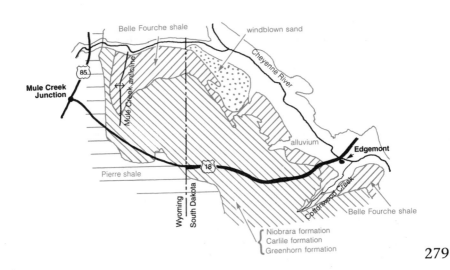

The bulldog fish, or xiphactinus, is ancestral to modern trout and salmon. This 13-foot specimen was collected 6 miles east of Hot Springs. —S.D. Museum of Geology

Fall River Falls

Just south of the road at the mouth of the canyon, Fall River tumbles over an outcrop of Fall River sandstone and travertine in a drop of about 50 feet. Travertine is a form of limestone that consists of calcite deposited directly from water solution. Most people know it best as the stuff that makes stalactites and stalagmites in caves.

In 1907 the city of Hot Springs built a low dam above the falls and carried the flow 4,700 feet through a flume to a point below the falls, where the water dropped 115 feet through a penstock to a small hydroelectric plant capable of producing 250 kilowatts. It supplied part of Hot Springs' electric power until the late 1960s.

Fall River Canyon

N. H. Darton called the Fall River formation the Dakota sandstone when he started his study of the rocks in the Black Hills around 1900. It was renamed the Fall River formation in 1927 after a study of its fossil plants revealed that the rock is much older than the original Dakota sandstone in eastern Nebraska.

The homogeneity and accessibility of the Fall River sandstone led to development of a building stone industry by 1890. Vertical sandstone faces near the mouth of the canyon on both sides of the road are the sites of those quarries. Other quarries around the Black Hills also produced the Cretaceous sandstones for construction material, but the quarries near Hot Springs were best equipped to cut stone

Quarry in the Fall River sandstone in 1936 and now. —S.D. Geological Survey

into sawn blocks. You can see sawn and rough blocks in the fine old sandstone buildings in downtown Hot Springs.

The excessive width of the sandstone hogback here is due both to exceptional thickness of the sandstones and to the Dudley anticline, which crosses the canyon on a north trend. The gently arching fold essentially doubles the width of the sandstone outcrop.

Outcrops along the highway display well the resistant Cretaceous sandstones, but poorly expose the much softer Jurassic Sundance, Unkpapa, and Morrison formations at road level. A soft lavender sandstone cropping out on the north side of the road on the

281

extreme eastern edge of town is the local Unkpapa sandstone. A geologic section measured along the north wall of the canyon shows the following between the mouth of the canyon and Hot Springs:

AGE	FORMATION	THICKNESS
CRETACEOUS		
	Skull Creek formation: dark gray-to-black marine shale	250 feet
	Fall River formation: tan-to-pink fine sandstone	147 feet
	Lakota formation: Fuson member: variegated clay and sandstone	123 feet
	Minnewast member: gray-to-tan limestone	29 feet
	Chilson member: buff-to-red sandstone	310 feet
JURASSIC		
	Morrison formation: grayish green shale	10 feet
	Unkpapa formation: variegated sandstone	140 feet
	Sundance formation: green shale and yellow sandstone	300 feet
TRIASSIC		
	Spearfish formation: redbeds and white gypsum	350 feet

Hot Springs

Hot Springs contains four items of special geologic interest: the old sandstone buildings, the cemented conglomerate along the canyon walls, the hot springs, and the mammoth site. In the downtown district, a cemented boulder conglomerate lines sections of the canyon walls to a height of several tens of feet. Evidently the canyon choked with gravel, probably in late Pleistocene time, then calcium carbonate, precipitated from the warm spring water, cemented the boulders into solid conglomerate rock. Fall River has since entrenched its channel, exposing the conglomerate in the valley walls.

A series of warm springs along Fall River encouraged development of a spa in the 1890s. Several hotels and bathhouses, ranging from modest frame structures to the elaborate Evans Hotel, encouraged people to come and "take the waters." They stressed the medicinal merit of their lithium content, which may even have helped a few manic-depressives. The railroad spur reached the springs in 1890. Spas are now out of vogue; the only spring still in use is Evans Plunge, which flows 9,000 gallons of water per minute at a lukewarm 89 degrees Fahrenheit.

Mammoth Site

In 1974 a contractor leveling an area north of the bypass for a housing development uncovered bones of the Columbian mammoth. When the abundance of bones showed that many skeletons were buried there, the contractor relinquished the site. A new nonprofit

Excavating mammoth site.
—S.D. Department of Tourism

corporation built an attractive building over the entire site, large enough to permit public viewing of the excavations in progress, and included a laboratory and storage facilities.

Excavation showed that the bones accumulated in a steep sinkhole created by solution and collapse of the underlying beds of gypsum. A sharp color change is obvious between the Spearfish redbeds and the tan-to-gray sinkhole filling, which consists of sand derived from the Sundance and Unkpapa formations. The filling is about 140 by 170 feet across at the level of the bone bed. Although the top of the sinkhole filling has been lost to erosion, current excavations and a drill hole show that at least 65 feet of bone-bearing fill is present.

Water rising through the sand would have kept it in partial suspension, making a quicksand. Smaller animals descended into the sinkhole and drank with impunity, but the enormous mammoths bogged down and died of exhaustion during their struggle with the quicksand. This condition may have persisted for between 400 and 1,000 years. By 1993, the skeletons of two woolly and 46 Columbian mammoths had been unearthed. Not all skeletons are complete, but all that include pelvises have been identified as males. Bones of many small species occasionally show up; the largest is the jaw of a short-nosed bear. Radiocarbon dates show that the bone bed is no more than 26,000 years old.

Cascade Springs and Falls

At Cascade Springs, 8 miles to the south on SD 71, artesian water rises to the surface through fractures near the crest of the sharply folded Cascade anticline. The flow is about 23 cubic feet per second,

Cascade Falls. Cascade Creek tumbles over a mass of travertine about a mile below Cascade Springs.

or 10,000 gallons per minute, with very little seasonal change. The water temperature is 68 degrees Fahrenheit. Chemical analysis of the water suggests that most of the flow is from the Minnelusa aquifer, the rest perhaps from the deeper Madison limestone. The low elevation of the spring influences the water level in both formations over a wide area.

Calcium carbonate, precipitated from the spring water, has built deposits of travertine along Cascade Creek. A mile and a half below the springs, the creek cascades over a mass of this travertine at Cascade Falls. Steeply tilted layers of Cretaceous sandstone are nicely exposed between the springs and the falls.

Between Hot Springs and Edgemont

A short distance west of Hot Springs, the highway climbs abruptly over the Cascade anticline, a sharp fold with the layers of rock most steeply tilted on its west side. The softer Spearfish redbeds eroded, so the route over the ridge is mostly built on the tilted layers of resistant Minnekahta limestone. A small outcrop of lavender shale near the crest of the fold gives a glimpse of the underlying Opeche shale. Geologists can trace the anticline north only a short distance into Paleozoic rocks, but can follow it south nearly to Nebraska.

Box folds in thin-bedded Minnekahta limestone. The folds probably formed during doming of the Black Hills.
—R. W. Schnabel, U.S. Geological Survey

The Red Valley is very wide between the area west of the Cascade structure and the old railroad junction of Minnekahta. To the north are long gentle dip slopes of Minnekahta limestone; to the south is the main hogback with its steep inner slope of Jurassic sandstones and shales and crest of Lakota sandstone.

Cottonwood Springs Dam was built where that creek cuts south through the Minnekahta limestone outcrop. It is strictly a flood-control structure that normally impounds little water. Minnelusa sandstone underlies most of the impoundment area.

South of the Minnekahta junction, the road parallels the axis of the erosionally breached Chilson anticline, a sister to the Cascade anticline. Erosion commonly hollows out the crests of anticlines, converting them into valleys. That happens because the rocks fracture as they stretch across the top of the fold. The fractures admit water, which promotes weathering in the rock, making it vulnerable to erosion.

The road swings to the southwest, climbs a long slope of Jurassic and Cretaceous sandstones, crosses a high bridge over Red Canyon at the crest of the road and descends several miles nearly to Edgemont through sandstone dip slopes.

Cycads

Cycads are primitive plants, the precursors to the grasses. They superficially resemble palms but actually are more closely related to the ferns. A cycad trunk stripped of its fronds somewhat resembles a pineapple or cluster of pineapples. Cycads appeared during Permian time, about 225 million years ago, and thrived mightily during Jurassic time. They still abounded in early Cretaceous time, but dwindled with the appearance of flowering plants. Cycads have been rare since then.

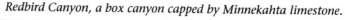

Redbird Canyon, a box canyon capped by Minnekahta limestone.

Fossil cycad, 30 inches high, from the vicinity of the former Cycad National Monument. It is now on display at the School of Mines.

About 2 miles west of Parker Peak, along the present route of US 18, someone discovered many well-preserved petrified cycads in 1892. Professor Ward of Yale University collected and shipped them back to New Haven for study. The National Park Service set aside an area of about a half square mile as Cycad National Monument in 1922, but the collecting had been so thorough that few cycad fragments remained. The monument was never developed, and it was withdrawn from the system in 1952. When US 18 was rerouted to its present location in 1981 and 1982, many fine cycad specimens appeared. You can see some of them in the Museum of Geology at the School of Mines in Rapid City.

Uranium

In 1951, showings of the uranium oxide mineral carnotite were identified in the Lakota sandstone in Craven Canyon, north of Edgemont. The usual flurry of leasing and exploration drilling followed, and many small prospects were opened.

Further development revealed that the ore occurred in channel sandstones within both the Fall River and Lakota formations. Many uranium minerals were later identified. The principal yellow ore minerals were carnotite, tyuyamunite, and metatyuyamunite; the dominant black minerals were uraninite, coffinite, paramontroseite, and haggite.

287

Crumpled cross-bedding in lower Cretaceous sandstone near Edgemont.
—W. W. Rubey, U.S. Geological Survey

The first shipments of ore were sent to Colorado for processing. A company completed a mill at Edgemont in July 1956, then later expanded it to recover the associated vanadium. The mill operated until 1973. Nothing has happened since the mill and mining properties were sold to the Tennessee Valley Authority in 1974. None of the old mines are visible from the highway.

Grindstones

A stone quarry that opened about 1895 just north of Edgemont supplied the building stone used in the older buildings. One bed within the Fall River formation consisted of fine sand grains that were so uniform in size and so sharply angular that the rock was good raw material for grindstones. Quarrying started about 1900. Rough blanks were hand dressed, then shaped on a water-driven lathe. A large load of finished stones was shipped east, where it found a ready market. However, the distance from market and the increasing popularity of carborundum abrasives soon doomed the project.

Between Edgemont and the Wyoming Line

West of Edgemont, the highway climbs out of the Cheyenne River Valley. It crosses the gently arched Cottonwood anticline, which trends southwest. Outcrops of the slabby Greenhorn limestone partly outline the fold.

Mainly, the highway follows the outcrop of the Carlile shale, with a few swings onto the Niobrara chalk. The chalk is light bluish gray where fresh, pale buff to yellow where weathered. The Carlile shale is a dark gray shale deposited in shallow seawater during late Cretaceous time. It contains the Turner sandy zone and three zones of large limestone concretions. Watch for these zones below road level at two or three points north of the highway. These zones of concretions continue west into Wyoming and northwest into Montana, where they appear in the Frontier formation.

The low and discontinuous ridge south of the highway exposes the Sharon Springs member of the Pierre formation, which contains many thick bentonite beds. Here and there it also contains well-preserved bones. Paleontologists have collected swimming marine reptiles and pterodactyls. A short distance west of the state line, the road shifts onto the Pierre shale, but outcrops are poor.

US 85
Wyoming Line–Spearfish
38 miles

The highway enters the state on the Minnelusa formation, dipping gently to the west. It soon crosses the major Fanny Peak monocline, which raises the eastern Black Hills fault block several hundred feet above the less-elevated Bear Lodge block. At O'Neil Pass, the road starts down the geologic section to the upper Cambrian rocks in Spearfish Canyon, back up to the Mississippian limestone near the ski areas, then rapidly down through the Paleozoic sediments to the Lead dome. There, ancient schists and lower Paleozoic rocks are cut and mineralized by intrusions of Tertiary rhyolite. From Deadwood to Spearfish, the route gradually climbs from the Cambrian rocks at Lead to the Permian-Triassic redbeds at Spearfish.

Between the Wyoming Line and Deadwood

Between Buckhorn, Wyoming, and a point about a mile inside the South Dakota border, the highway crosses poorly exposed sandstone and limestone of the Pennsylvanian and Permian Minnelusa formation.

The road turns north for 2 miles, parallel to the western flank of the Fanny Peak monocline. This fold divides the Black Hills uplift into a western block that includes the Bear Lodge Mountains of

Wyoming, and an eastern block that encompasses the Black Hills. The flexure closely follows the boundary between South Dakota and Wyoming. The strata in the monocline tilt down to the west at about 12 to 20 degrees, but probably steepen and become a fault with depth, as in most monoclines. Near the road, the western block dropped 500 feet across the monocline. South of Newcastle, Wyoming, the displacement becomes much greater, and the monocline becomes the eastern margin of the deep Powder River basin.

O'Neil Pass, 6,690 feet high, is the drainage divide between Sand Creek and Spearfish Creek, two of the major streams that flow north out of the northern end of the Black Hills. Bedrock is the Minnelusa formation.

Wyoming line to Spearfish (38 miles).

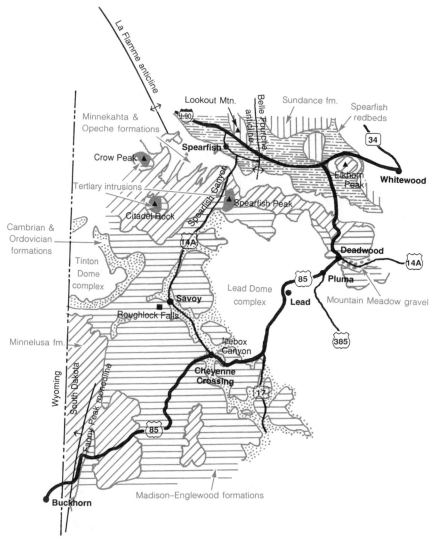

Limestone cliffs form the upper end of Spearfish Canyon, east of O'Neil Pass.

About a mile east of O'Neil Pass, the highway enters the headwaters of Little Spearfish Canyon and proceeds between canyon walls of Madison limestone to Cheyenne Crossing, where Little Spearfish Creek joins the main creek. This beautiful, undisturbed area has been set aside as a watershed to supply part of the water for Lead, Deadwood, and the Homestake Mine. For 2 or 3 miles on either side of the junction, the creeks have eroded their canyons into late Cambrian and Ordovician formations.

East of Cheyenne Crossing, the highway climbs through Icebox Canyon, aptly named. At Cheyenne Crossing, the lower canyon walls expose most of the Cambrian Deadwood formation. About halfway through the canyon, the highway crosses poorly exposed outcrops of green shale in the Ordovician Icebox formation, reddish tan dolomite in the Whitewood formation, and thin beds of pink dolomite in the Englewood formation. Above those formations rise the gray walls of Madison limestone.

At the drainage divide at the head of Icebox Canyon, County Road 194 goes north past Deer Mountain and ends at the lookout atop Terry Peak. A conical mass of igneous rock at least 300 feet thick caps Terry Peak. The mass intruded the Deadwood formation during Tertiary time and what we now see is probably the eroded core of a laccolith. The igneous rock in Deer Mountain is probably another denuded laccolith, though much smaller and not nearly as high.

Like all laccoliths, these took shape as blisters of molten magma intruded between layers of sedimentary rock, bulging them up like a mushroom cap.

The highway drops slowly into the drainage of Whitetail Creek. The roadside rocks are a complex of igneous intrusions emplaced within the shales of the lower Deadwood formation, which overlie steeply inclined Precambrian schists. On the upland west of the road and south of Nevada Gulch, the new Golden Reward Mining Company produces gold ore from the Deadwood formation. The original Golden Reward Mining and Milling Company of a century ago found most of this ore far too lean to mine. The rusty schists on the southwestern edge of Lead are in the Homestake formation, but contain no gold.

The Lead-Deadwood Dome

Tremendous volumes of molten magma were forced upward into the earth's crust in the Lead and Deadwood area roughly 50 million years ago. The magma rose as thin sheets or small pipes, following steeply inclined planes of weakness in the ancient metamorphic rocks. Intrusion of these dikes swelled the volume of the rocks, bulging the old erosion surface that separates the Precambrian and Cambrian rocks and doming the overlying Paleozoic rocks. Some of the magma rose into the overlying sediments, where it formed either bulging laccoliths or flat sills, thus further accentuating the dome in the higher rocks. In turn, the raised sedimentary rocks were rapidly eroded, exposing the older intricately folded and intruded rocks. The area where the Paleozoic rocks have been largely removed is called the Lead-Deadwood dome. Without the exposure of deeper rocks in the dome, the huge Homestake gold deposit would not have been discovered.

Lead grew up around the many early mines that worked the Homestake lode. In the 1930s, the old business district, which was

Idealized cross section of the Lead-Deadwood dome.

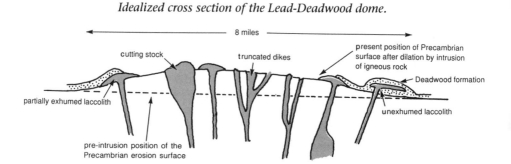

8 miles

cutting stock

truncated dikes

present position of Precambrian surface after dilation by intrusion of igneous rock

Deadwood formation

partially exhumed laccolith

unexhumed laccolith

pre-intrusion position of the Precambrian erosion surface

292

farther down the hill opposite the open cut, started to collapse into the early underground workings. The business district promptly moved up the hill to its present location. The early miners left so much rich ore near the surface that the Homestake Mining Company is again moving parts of the town to extend the open cut farther south. The red and greenish gray schists exposed in the open cut are part of the Homestake, Poorman, and Ellison formations. Pale yellow dikes of much younger Tertiary rhyolite cut both formations. The road passes below the mills to the junction of US 85 and US 385 at Pluma, then follows Whitewood Creek to Deadwood.

Between Deadwood and Spearfish

This stretch is one of the best places in the Black Hills to see the complete section of Paleozoic and Triassic sedimentary rocks. The following formations are exposed, from the schist at Deadwood to the red shale at Spearfish:

AGE	FORMATION	THICKNESS
PERMIAN-TRIASSIC		
	Spearfish redbeds	700 feet
PERMIAN		
	Minnekahta limestone	35 feet
	Opeche redbeds	110 feet
PENNSYLVANIAN and PERMIAN		
	Minnelusa formation	500 feet
MISSISSIPPIAN		
	Pahasapa, or Madison, limestone	550 feet
	Englewood formation	70 feet
ORDOVICIAN		
	Whitewood dolomite	40 feet
	Roughlock siltstone	30 feet
	Winnipeg shale	70 feet
CAMBRIAN		
	Deadwood formation	400 feet
PRECAMBRIAN		
	schists and other metamorphic rocks	

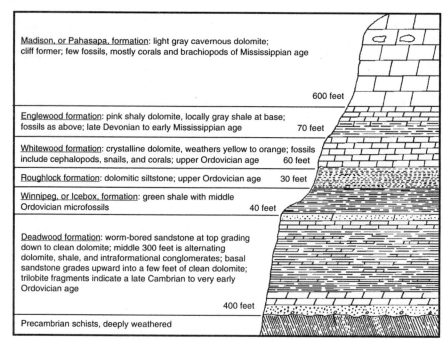

Madison, or Pahasapa, formation: light gray cavernous dolomite; cliff former; few fossils, mostly corals and brachiopods of Mississippian age

600 feet

Englewood formation: pink shaly dolomite, locally gray shale at base; fossils as above; late Devonian to early Mississippian age   70 feet

Whitewood formation: crystalline dolomite, weathers yellow to orange; fossils include cephalopods, snails, and corals; upper Ordovician age   60 feet

Roughlock formation: dolomitic siltstone; upper Ordovician age   30 feet

Winnipeg, or Icebox, formation: green shale with middle Ordovician microfossils   40 feet

Deadwood formation: worm-bored sandstone at top grading down to clean dolomite; middle 300 feet is alternating dolomite, shale, and intraformational conglomerates; basal sandstone grades upward into a few feet of clean dolomite; trilobite fragments indicate a late Cambrian to very early Ordovician age

400 feet

Precambrian schists, deeply weathered

Formations exposed at Deadwood.

In the northeastern part of Lead, the reddish brown Deadwood formation is exposed on both sides of the canyon. On the east side, across the creek, the basal sandstones of the Deadwood formation rest on deeply weathered, steeply dipping schists. Waterworn cobbles of white quartz lie at the base of the Deadwood formation, directly on the old erosion surface. Above, beach sands grade upward to shales deposited as the sea deepened. On the west side of the highway is a towering cliff of reddish brown Deadwood sandstones, shales, and limestone conglomerates, 400 feet of them. This outcrop is the thickest and best-exposed section of the Cambrian Deadwood formation, the designated type section for comparison with other Deadwood outcrops in the Black Hills.

At the north end of town, where US 85 and US 14A divide, the Scolithus sandstone at the top of the Deadwood formation has dipped down to road level. The Scolithus fossils appear to be worm borings. They are not much to look at, but their Cambrian antiquity does lend them a certain dignity.

Above the Scolithus horizon, and slumping down over it, are about 60 feet of green Winnipeg shale. Above that is a silty and slabby limestone about 30 feet thick, the Roughlock formation. This in turn grades upward into about 40 feet of clean, orange-to-brown White-wood dolomite, named for its exposures here along Whitewood Creek. It contains many fossils of late Ordovician age, including several varieties of corals, snails, and cephalopods. The Whitewood formation is continuous underground with the Red River dolomite in Manitoba and the Bighorn dolomite in Wyoming and Montana. The upper part of the formation produces oil in the northwestern corner of the state.

North of its junction with US 14A, US 85 climbs across the remainder of the Madison limestone. Then, for several miles it follows the canyon of Polo Creek, passing good exposures of Minnelusa sandstones and dolomites. Near the mouth of the canyon, the Minnelusa formation is overlain by about 110 feet of red silty Opeche shale, capped by a conspicuous ledge about 30 feet high of pink-to-red Minnekahta limestone. A quarry at the mouth of the canyon furnishes crushed limestone for concrete aggregate and road metal.

Deadwood formation dipping off the northeast flank of the Black Hills at the lower end of Deadwood.

US 385
Hot Springs–Pluma
88 miles

US 385 traverses the Black Hills from south to north, crossing a wide variety of rock formations. Except at the two ends of the traverse, the rocks are all very old, none less than 1.7 billion years. Nearly all were deposited as muds and sands in ancient seas, some in shallow water, but most in very deep water. Then they were deeply buried and recrystallized at high temperature to become metamorphic rocks, mostly schists.

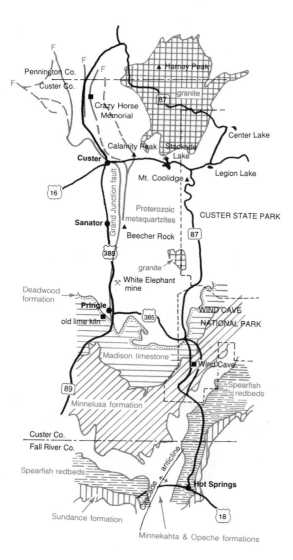

Hot Springs to Pennington County line (39 miles).
F = *fault.*

Watch between Pringle and Pluma to see how the highway engineers have designed their roadcuts according to the layering in the schist. Where the planes of weakness dip toward the road, they took advantage of them to make a smooth cut, even though it required removing more rock to allow stability. Where the layering dips away from the road, the cuts are jagged. Where the layering in the rocks is nearly horizontal, it is often possible to make nearly vertical cuts.

Hot Springs

The warm springs at Hot Springs attracted successive Indian tribes as settlers forced them west. Then the settlers also recognized the potential of the springs. Spas where wealthy people could "take the waters" to cure their ills were the rage in Europe and the eastern United States. Settlers soon built several wood-frame bathhouses and hotels along Fall River, and the Chicago and Northwestern Railroad built a spur line from Buffalo Gap in 1890.

More imposing buildings of native sandstone slowly replaced the wooden structures. The high point in the development of Hot Springs came in 1892 with completion of the magnificent Evans Hotel. Spas declined in popularity after World War I, and Hot Springs became a medical center, complete with hospitals, sanatoriums, and retirement facilities such as the State Soldiers' Home.

Fall River cut its deep canyon through sedimentary rocks that dip steeply southeast, exposing a section through a series of formations that represent quite a long span of time. The oldest is the Minnelusa formation, which appears at the northwest edge of Hot Springs. Below are the formations exposed in the canyon walls in town:

FORMATION	ROCK TYPES	THICKNESS
TRIASSIC		
	Spearfish formation: redbeds and white gypsum,	up to 350 feet
PERMIAN		
	Minnekahta formation: thin-bedded pink limestone	120 feet
	Opeche formation: red siltstone and shale	40 feet
	Minnelusa formation: sandstone and dolomite,	up to 400 feet

The older part of Hot Springs is on the valley floor, on red shale and siltstone of the Spearfish formation. Residential areas on both sides are on stream terraces as much as 250 feet above the present Fall River. Red shale that contains beds of massive white gypsum, also in the Spearfish formation, is exposed in the northern part of town. Watch for cliffs of red Minnekahta limestone beneath the redbeds near the springs in the northwestern part of town. A spectacular outcrop of the sandstones and dolomites of the Minnelusa formation is a short distance up the canyon west of Evans Plunge.

A cemented boulder conglomerate lines sections of the canyon walls in the downtown district to a height of several tens of feet. At some time, probably in late Pleistocene time, clay and gravel partly choked the canyon. Then calcium carbonate, precipitated from the warm spring water, cemented them into solid rock. Finally, the Fall River cut deeply into the conglomerate, leaving remnants exposed along the canyon walls.

The Springs

The springs produce from artesian aquifers or limestone caverns. The water falls as rain and snow on the higher areas to the north and northwest. It filters down to the groundwater table, circulates to great depths, where it absorbs heat from the earth, and returns to the surface where streams have cut deep canyons into the rocks. The calcium carbonate and calcium sulfate in the water was probably dissolved out of the Madison and Minnelusa formations. The water emerges through fissures in the bedrock. You can see them at some springs, but in most you see the water issuing from the gravel that covers the fissures.

People tend to attribute medicinal properties to springs in proportion to how nasty the water tastes and how foul it smells. The water at Hot Springs has no nasty taste or foul smell, so people advertising its medicinal value have emphasized other curative properties. Early advertisements praised its pleasant temperature of 89 degrees Fahrenheit, its trace of lithium, and its very mild laxative effect. The dissolved mineral content of about 1,500 parts per million is actually less than that of the municipal supply of many South Dakota towns that rely on deep artesian wells.

Gypsum

Outcrops of white gypsum in the form of granular alabaster are conspicuous in the northern part of Hot Springs. One bed of commercial quality reaches a thickness of 30 feet. The Hot Springs Plaster Company started a two-kettle, 60-ton mill in 1893 to produce wall and dental plaster; they did not rebuild after the mill burned in 1909.

Cold Brook Reservoir

A dam impounds Cold Brook where it cuts through the Minnekahta limestone just north of the city limits, about a mile west of US 385. The reservoir is used for flood control and recreation. Bedrock under the reservoir is highly permeable sandstone in the upper part of the Minnelusa formation. It was necessary to grout it under the dam to prevent excessive leakage.

Between Hot Springs and Pringle

The exposed rocks become progressively older, from the Triassic redbeds at Hot Springs to Precambrian schists and pegmatites at Pringle. Over much of the upland surface, a thin veneer of gravel conceals the truncated layers of Paleozoic rocks. The route crosses the redbeds nearly to the southern boundary of Wind Cave National Park, where the road drops onto the Permian Minnekahta limestone, then onto a surface underlain by upper Paleozoic limestones.

The log of a water well at the park gives an idea of the thicknesses of the formations the highway crosses between Pringle and Hot Springs. The well is in a canyon 2 miles east of the cave entrance, where probably 200 feet of the upper Minnelusa sandstone has been eroded. The well log shows the following formations and thicknesses:

AGE	FORMATION	THICKNESS
PERMIAN-TRIASSIC		
	Minnelusa sandstone and limestone, red and yellow	472 feet
MISSISSIPPIAN		
	Pahasapa, or Madison, dolomitic limestone, gray	258 feet
	Englewood limestone, pink	30 feet
CAMBRIAN		
	Deadwood sandstone, red and brown	25 feet
PRECAMBRIAN		
	schist, gray	10 feet

Wind Cave National Park

According to local tradition, a cowboy discovered Wind Cave in 1884, when he noticed air whistling out of an 8-by-10-inch opening in the Mississippian Madison limestone. The atmospheric pressure

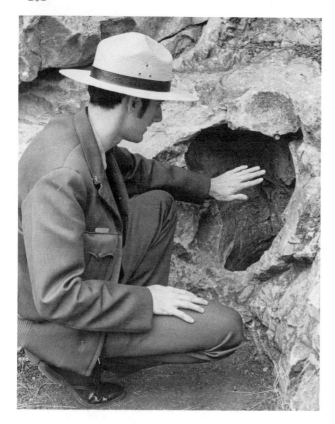

*Original entrance to
Wind Cave.* —Art Palmer

Chert ceiling in Wind Cave. —Art Palmer

must have been dropping. Shortly thereafter, two local ranchers filed claims on the surrounding area and later opened the cave to the public. Soon its features were thought worthy of national park status. The area was closed to entry in 1900, and Congress established the national park in 1903. The park also includes a large game preserve with buffalo, elk, deer, antelope, and other native animals.

Wind Cave is a typical limestone cavern, eroded as circulating groundwater dissolved holes in Mississippian Madison limestone. The irregular enlargement of vertical fractures created most of the openings, mainly along the dominant set of fractures, which trends southeast. Enlargement of those and of a lesser fracture sets at right angles to the main set gives a rude grid pattern to the map of cave openings. In some places, solution has extended laterally by etching out more vulnerable layers of rock. That produces cave rooms of considerable area, but no great height. The ceiling of one large room is a continuous layer of insoluble chert.

Wind Cave has most of the usual forms of cave ornamentation. Its unique feature is boxwork, which probably forms where two sets of closely spaced vertical fractures broke thin layers of limestone into small rectangular blocks. Calcite filled the fractures, and then

Upper Minnelusa sandstone along Beaver Creek in Wind Cave National Park. The sandstone was jumbled by dissolution of interbedded layers of gypsum. —C. L. Baker

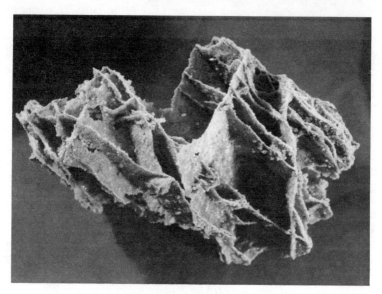

Detail of boxwork in Wind Cave. —S.D. Museum of Geology

the original limestone blocks dissolved, leaving only the delicate crystalline partitions. These are best seen in an area fittingly called the Post Office.

After the cave drained, water seeping from the walls and ceilings deposited stalactites, stalagmites, flowstone, and the delicate crystals called popcorn. Wind Cave is the only cavern in the Black Hills where cave explorers have been able to descend to the water table. Instruments there record the seasonal variations in the water level in the Madison limestone.

In 1931 Congress appropriated money to install a primitive electric lighting system. It grew until indirect lighting now illuminates all public areas. An elevator shaft was sunk in 1934–35 to a depth of 212 feet, with stations at 119 and 195 feet. It penetrated only 30 feet of solid limestone; the rest was honeycombed with large and small cave openings.

Between park headquarters and the western boundary of the park, the highway drops through a thin section of dark brown Deadwood sandstone and onto the Precambrian siliceous schists and quartzites cut by intrusive masses of resistant pegmatitic granite.

Pegmatite mining on a large scale started in the 1920s, and it continues on a small scale. You can see the dumps of several mines from the highway. The principal mineral sought is feldspar, used in ceramics, for glazes for plumbing fixtures, for abrasives, and in the manufacture of glass and fiberglass. Mica, beryl, and lithium minerals are important by-products.

Pringle

Pringle has been a small mining and mineral-processing center since the Burlington Railroad built its line through the center of the Black Hills in 1891. Local limestone was first quarried and burned to make lime in small beehive kilns built of masonry. In 1899, two larger commercial kilns replaced the earlier primitive rock structures, operating until about 1975. Much of the lime was shipped to the gold mills in Lead and Deadwood; some went to sugar beet factories in South Dakota, Nebraska, and Wyoming.

Fracturing Sand

Coarse sand in the lower part of the Deadwood formation was mined near Pringle for use in oil well drilling during the late 1960s and early 1970s. One way to increase oil production is to hydraulically fracture the reservoir rock by pumping water or oil into it at very high pressure. Well-rounded coarse sand mixed with the fluid enters the fractures and props them open after the pressure is released. Oil moves more freely through the fractured rock and into the well.

Between Pringle and Custer

The highway follows the broad valley of Beaver Creek, where erosion of the soft schists has left the pegmatite dikes standing out in bold relief.

Along most of the distance between Pringle and Custer, US 385 parallels the Grand Junction fault, a major structure. The fault breaks only very old rocks, and the fault plane was itself folded during emplacement of the Harney Peak granite and pegmatite. Here, rock on one side of the fault is impure sandstone; rock on the other side is black shale. All are equally vulnerable to weathering and erosion, so the landscape does not express the fault. Rocks farther east of the fault are largely quartzites deposited during early Proterozoic time. The Harney Peak granite intruded them.

White Elephant Mine

A large, elliptical mass of pegmatite high on the hillside east of the highway was for many years a source of feldspar that was almost free of iron. It was shipped to Illinois for use in glazing bathroom fixtures. The low iron content ensured a sparkling white glaze. The mine also produced small quantities of white and rose quartz, mica, and beryl, all of which were sold locally.

Beecher Rock

Two big blocks of granite appear on the skyline to the east, 0.75 mile southeast of Sanator. They are erosional remnants of a much longer dike. Custer's 1874 map identifies them as Turk's Head. They were later named for Henry Ward Beecher, controversial evangelist and brother of Harriet Beecher Stowe, who wrote *Uncle Tom's Cabin.*

The two spires of Beecher Rock.

Custer

Although placer gold was known in the Black Hills before the Civil War, their cession to the Sioux tribes in 1868 prevented mining. When General Custer "discovered" placer gold in the gravels along French Creek in 1874, he brought it to the attention of the newspapers. The army could do nothing to deter the miners who then swarmed into the Black Hills. They laid out Custer City in July 1875 in blatant defiance of the military ban.

Construction began in earnest in late fall. By March, the city had 1,400 buildings and a population of between 6,000 and 10,000 people. Then, the miners abruptly abandoned Custer as they swarmed to the new and much richer placer diggings in Deadwood Gulch. A few weeks later, Custer shrank to a population of 14. When the Black Hills were formally opened to white settlement in 1877, some miners and merchants, discouraged by the chaos at Deadwood, returned to Custer. By the summer of 1877, it had a stable population of about 500.

Cluster of hexagonal beryl crystals from a cavity in the Crown pegmatite mine northwest of Custer. —S.D. Museum of Geology

As they exhausted the placer deposits, the miners began to prospect for the bedrock source, the mother lode. They developed several small gold prospects along French Creek northwest of Custer, but they never found the main bedrock source of the placer gold. A mill at the south edge of town processes crude feldspar, lithium minerals, mica, beryl, and other rare pegmatite minerals. Forest products, pegmatite mineral specimens, and tourism are now the town's main industries.

Between Custer and Pluma

Between Custer and the junction with SD 44, the highway roughly follows the trace of the ancient Grand Junction fault, which separates two unrelated sequences of schist. One is thinly layered and contains crystals of red garnet; the other is thickly layered and contains only a few garnets. Several early precious-metal prospects were along this interval. The gold, associated with sulfides, occurs in quartz veins, along shear planes, and along the Grand·Junction fault.

Pegmatitic granites in masses of all sizes stand in high erosional relief above the less-resistant schists, which they intrude. The number and size of the pegmatite masses increase toward Harney Peak. A geologic map of the Berne quadrangle, northwest of Custer, covers 53 square miles. Geologists estimate that some 1,150 separate pegmatite bodies exist within that area, yet because of their small size they cover less than 1 percent of the land.

Bedrock between Hill City and SD 44 consists of metamorphosed sedimentary rocks, originally shales and muddy sandstones, now mostly schists. Ancient faults that trend generally north cut them.

Sheridan Lake

The Bureau of Reclamation built an earthen dam across Spring Creek in 1940. It is 850 feet long and stands 158 feet above bedrock. The lake covers 380 acres. The growth of cattails along the highway

Pennington County line to Pluma (49 miles).

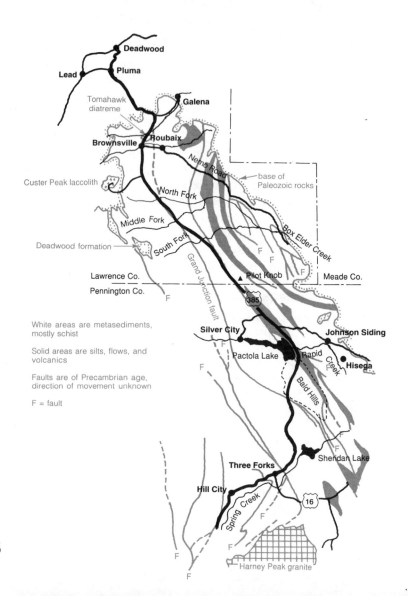

at the upper end of the reservoir shows that it is filling with sediment. A low horizontal grade just above the water line at the head of the cattail patch marks the point where an early flume diverted water from Spring Creek to the placer mines at Rockerville.

Calumet Peak

Miners from the Calumet copper district in northern Michigan named the high mountain southeast of Sheridan Lake Calumet Peak. The Jenney-Newton geological survey of the Black Hills found shows of copper there in 1875. The ore minerals are in a shear zone within a series of quartzites and slates that contain graphite. The metallic minerals include the iron sulfides pyrite and pyrrhotite and a little of the iron and copper sulfide, chalcopyrite. The latter weathered near the surface to produce colorful shows of copper in greenish blue malachite and intensely blue azurite.

Efforts to start a copper mine began in 1879. Early development included a shaft 850 feet deep and many hundreds of feet of lateral exploration drifts. Miners found no commercial body of copper ore, although the property did produce a small tonnage of low-grade ore during World War I. Ruins of a small brick smelter and an aerial tramway remain.

Pactola Dam and Reservoir

Pactola dam is an earthen structure across the valley of Rapid Creek. The Bureau of Reclamation built it to provide flood control and recreation and to store water for irrigation and domestic use. At normal operating level, 41 feet below the spillway, it stores 55,000 acre-feet of water. The rocks next to the dam are metamorphosed gabbros; at the spillway are basalt lava flows.

Two roadcuts east of the highway, 0.8 mile north of the spillway, expose the only known example in the Black Hills of a fault that moved on a nearly horizontal surface. Black basalt above the fault lies on thinly layered black cherty shale below. No one knows in what direction the fault moved, or how far.

Silver City

County Road 141 leads west to Silver City, which is at the head of Pactola Reservoir. This was a center of placer mining in the early days and again in the depression years of the 1930s. About 1900, exploration north and west of town centered on several quartz veins that contain the lead and antimony sulfide mineral boulangerite, as well as varying amounts of gold and silver. The small volume of ore and its low grade discouraged mining.

Box Elder Drainage

The sharp peak of Pilot Knob rises 300 feet in the area east of the highway, 0.25 mile north of Jim Creek. Between Pilot Knob and Custer Peak, the road successively crosses three forks of Box Elder Creek. Much of the bedrock is slaty and graphitic shale. The very wide valley of the Middle Fork, known as Mountain Meadow, gives its name to the old land surface that formed as Oligocene sediments of the White River group partially buried the Black Hills. Roubaix Lake is a small impoundment on the Middle Fork about 1 mile west of US 385.

Custer Peak

Custer Peak is a conspicuous landmark in the northern Black Hills, 2 miles west of US 385. Its symmetrically conical form probably explains why some early travelers in the Black Hills reported seeing extinct volcanoes. Although Custer Peak is not a volcano, it does consist of igneous rock.

When the Black Hills were rising, still with a thick blanket of sedimentary rocks covering their central part, the molten magma that forms Custer Peak rose through the Precambrian schists with their steeply inclined layers. When the magma reached the nearly horizontal layers of the Deadwood sandstone, it spread within the weak shales in the middle of the formation to make an irregular blister of molten rock nearly a mile and a half across and 800 feet thick at the center. The magma crystallized into igneous rock there, and erosion later stripped off its cover of less-resistant sedimentary rocks. Now you see that blister of igneous rock as Custer Peak, a cone of rhyolite standing on nearly horizontal sandstone in the lower part of the Deadwood formation.

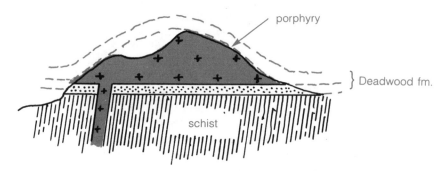

Section through Custer Peak.

Adit of an abandoned mine in the northern Black Hills. Enter such places only with permission and take safety precautions.

Brownsville

Brownsville was originally a station on the Black Hills and Fort Pierre Railroad, which was built by Homestake Mining Company before 1900. Like other narrow-gauge routes in the Black Hills, it hauled timber and supplies to the mines and hauled ore from the mines to the mills in Deadwood. The line was abandoned in 1930. East of Brownsville 1.5 miles lies the settlement of Roubaix, site of the abandoned Cloverleaf, or Uncle Sam, gold mine.

Tomahawk Diatreme

For about two-thirds of a mile north from Brownsville, embracing the upper end of the golf course, the highway crosses a wild assortment of igneous rocks that include bedded volcanic ash and assorted rubble in an area about a mile across. This is a diatreme, and it formed explosively as molten magma came to the surface in a great blast of escaping gas, probably mostly carbon dioxide and steam.

The first roadcut on the northeast corner of the junction of Nemo Road and US 385 exposes a block of Deadwood shale that collapsed into the cavity that had been opened by the stream of escaping gas and magma. The block is in vertical contact with volcanic ash.

The best exposure of the volcanic rocks is in a small knoll behind the Tomahawk clubhouse. Fragments of Cretaceous and lower Paleozoic rocks have been found in the rubble. The Cretaceous rocks must have fallen as much as 3,500 feet to reach their present position. All were whirling around in the blast of escaping carbon dioxide and steam. The presence of the Cretaceous rocks in the diatreme leaves no doubt that Cretaceous rocks were above this area when it was emplaced and have since been lost to erosion. A ridge west of the highway, opposite the golf course, is part of another dropped block that contains sedimentary rocks ranging in age from Cambrian to Mississippian. Several age dates on the igneous rocks show that they arrived about 56 million years ago. Several dozen diatremes of similar age exist in central Montana. This one seems to be an eastern outpost of that activity.

Galena

Galena, an old silver-mining camp, is 3 miles east of US 385 on Bear Butte Creek. It was very active during the 1870s, but lawsuits and a drop in the price of silver shut it down in 1883. Several attempts to resume mining failed.

Mineralizing solutions emanating from the Tertiary igneous intrusions rose along vertical fractures to invade the Deadwood formation at two levels. The lower contact was a zone of porous sand and dolomite just above the basal quartzite. The upper contact was a similar sand and dolomite sequence below the upper quartzite of the Deadwood formation. Ore bodies were long shoestrings that followed the intersection of the contact zones and the vertical feeder fracture.

The primary ore minerals were galena and pyrite. Galena is a lead sulfide mineral. In these ores, it contained large amounts of silver as an impurity. In many of the mines, the ore was oxidized into a soft yellow material with the consistency of clay. Some of the early ore was so rich that miners could profitably haul it to the Missouri River by ox team for shipment by steamboat to the smelter at Omaha, Nebraska. They treated some of the leaner ore in a small smelter at Galena. A small slag pile marks the site.

Strawberry Ridge

The high divide between the Bear Butte Creek drainage at Galena and the Whitewood Creek drainage at Pluma is known as Strawberry Ridge. Most of the rocks are garnet schists of the Precambrian Grizzly formation, with Deadwood sandstone and shale capping the highest points. Many dikes and sills of finely crystalline igneous rock of early Tertiary age intrude both sedimentary rocks.

orht

Iron Mine

One mile south of the crest of Strawberry Ridge, nearly pure hematite occurs in the base of the Deadwood formation beneath a ledge of massive quartzite. The South Dakota State Cement Plant at Rapid City used the material in 1948 as a source of iron. The mined area is west of the road.

Pluma

Before World War I, Pluma was the site of a large custom mill that processed gold ores. You can see some of the elaborate rock foundations south of the highway, just east of the road junction. The plant originally used the chlorination process, then converted in 1903 to the cyanide leaching process. After the plant burned in 1912, the operation was rebuilt closer to the mines around Terry Peak.

SD 44
Rapid City–US 385
18 miles

Rocks near Rapid City are as young as Cretaceous; those at the other end of the route are Precambrian. The road passes a wide variety of interesting sites as it crosses through that vast abyss of time.

Cement in South Dakota

South Dakota is unique among states in owning a commercial facility for the manufacture of portland cement. The plant, on the west side of Rapid City, markets its product over a trade area roughly

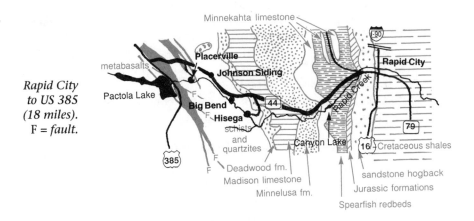

Rapid City to US 385 (18 miles). F = fault.

311

South Dakota State Cement Plant at Rapid City.

500 miles in radius. The story begins with the Western Portland Cement Company, which made cement at Yankton between 1891 and 1910. The raw materials, Niobrara chalk and Pierre shale, were quarried just west of town. Difficulty in maintaining a uniform product and competition from manufacturers farther east forced the plant to close in 1910 after producing nearly 2 million barrels of cement.

Two or three years earlier, anticipating the closure of the Yankton operation, the State School of Mines undertook a detailed study of the raw materials in the Black Hills. A superior product could be made from Minnekahta limestone and Cretaceous shale. Coal was available in nearby Wyoming, and a promise of natural gas from eastern Montana ensured cheap fuel.

When the Yankton plant closed, citizens felt that the commercial plants to the east, having no competition in this market, were taking advantage of the situation. Agitation started for construction of a state plant. Rivalry developed between Chamberlain, which wanted a plant using the same less-satisfactory raw materials as at Yankton, and Rapid City, which could document its claim to superior raw materials and inexpensive fuels. In February 1917, the state legislature authorized $1 million for a plant, and a commission later chose Rapid City as the site. Wartime shortages delayed the start of construction until 1923, but the plant was completed in late 1924, and the first cement was shipped early in 1925. The plant has been updated and its capacity increased as market conditions have dictated. The present capacity is 750,000 tons per year.

Between Rapid City and Canyon Lake

SD 44 follows the floodplain of Rapid Creek through its narrow water gap, then crosses the Red Valley. The water gap slices through the hogback ridge that divides eastern Rapid City from western Rapid

City. Shown below are the geologic formations exposed in the gap from east to west:

AGE	FORMATION	THICKNESS
CRETACEOUS		
	Belle Fourche shale	150 feet
	Newcastle sandstone channel, up to	35 feet
	Skull Creek shale	275 feet
	Fall River sandstone	125 feet
	Lakota formation:	
	Fuson member, variegated shale	110 feet
	Chilson member, sandstone	125 feet
JURASSIC		
	Morrison formation, shale	75 feet
	Unkpapa formation, white sandstone	125 feet
	Sundance formation, sand and shale	450 feet
PERMIAN-TRIASSIC		
	Spearfish formation, redbeds and gypsum	375 feet

Some of these formations are hard to distinguish from road level. For a better view, drive south on West Boulevard to Quincy Street, then west up the east flank of the hogback through the sandstone section. Stop at Hangman's Rock and walk to the viewpoint a few feet to the north. If you look north across the gap, you will see the

Fall River sandstone on north side of the water gap at Rapid City. It forms an important aquifer under the surrounding prairie.

Channel of Newcastle sandstone cut into the Skull Creek shale north of the water gap in Rapid City. It holds up the crest of the hill.

white Unkpapa sandstone to the left, the Lakota sandstone where the M is, and the Fall River formation in the sandstone ridge behind the packing plant. An interrupted ridge with trees on it farther east and north is Newcastle sandstone, filling an old stream channel. The view from the dinosaur concession is even better.

Why the dinosaurs on Skyline Drive? The Morrison shale crops out just beneath the Lakota sandstone on the ridge crest. Throughout this area, both formations carry dinosaur fossils. Fifty years ago it was possible to pick up fragments of dinosaur bone within 100 feet of the big concrete brontosaurus. Craftsmen of the Works Progress Administration built the dinosaurs during the Great Depression of the 1930s. They scaled them up from small models created in the American Museum of Natural History.

The Flood of 1972

The greenway along Rapid Creek between the east edge of town and the golf course just below Canyon Lake is in the area devastated by a disastrous flood on the night of June 9, 1972. As much as 11 inches of rain had fallen in the drainage of Rapid Creek below the Pactola Dam in the hours before the flood. Rapid Creek reached a peak discharge of about 50,000 cubic feet per second.

The flood washed away or damaged hundreds of buildings and killed 268 people. The storm affected all drainages from Sturgis to Keystone, but property damage and loss of life were very small along the other streams because fewer people lived there.

Canyon Lake Dam

The original dam on Rapid Creek was a log structure built about 1885 where the creek cut a narrow gorge through the resistant Minnekahta limestone. Irrigators from downstream dynamited the structure in 1900 because they believed that too much water was evaporating. The present earthen dam was built at the same location in the early 1930s. After it was breached in the 1972 flood, it was rebuilt to impound a substantially smaller reservoir.

Three large springs surface just above Canyon Lake. The city water system uses two, and the third furnishes water to the Cleghorn Springs State Fish Hatchery. The discharge is hard to measure, approximately 13 million gallons per day.

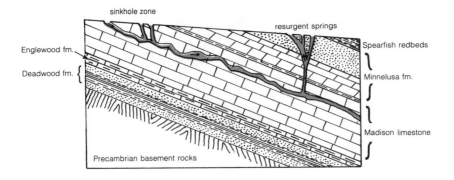

Profile along the bed of Rapid Creek above Canyon Lake. Water entering sinkholes reappears in resurgent springs in the Minnelusa sandstone.

Look just above Canyon Lake, on the north canyon wall behind the church, for good exposures of the Minnekahta limestone, the red Opeche shale, and the uppermost yellow Minnelusa sandstone. Another cliff, a half mile upstream, exhibits complex structures that formed when rocks collapsed down into caverns of dissolved gypsum and anhydrite beds in the middle part of the Minnelusa formation. Some of this solution happened soon after deposition, but most of it occurred after Eocene time when the Black Hills rose and were deeply eroded.

Between Canyon Lake and US 385

Where the highway leaves the valley bottom and abruptly climbs 700 feet to the old upland erosion surface, it crosses a long section of tilted dolomite beds in the lower part of the Minnelusa formation. Watch along this stretch for a short flat interval in which the road follows an old river terrace; it probably corresponds to the Rapid terrace farther east. The glimpse of red clay in a low roadcut opposite the former site of Potter's sawmill marks the contact between the Minnelusa formation and the underlying buried erosion surface on the Madison limestone.

Mississippian coral from the Madison limestone found in terrace gravel above Rapid Creek. —S.D. Museum of Geology

Two commercial caves in the Pahasapa, or Madison, limestone are near the highway; their entrances are about 2,300 feet apart. The lower cave contains dolomite sand that consists of tiny crystalline rhombs that were freed as the more soluble calcium carbonate that enclosed them dissolved.

Between the sawmill and the Hisega turnoff, the road crosses the Madison limestone or high terrace gravels covering it. Two short pale gray roadcuts are typical of the part of the Madison limestone that contains a lot of dolomite. The underlying Englewood limestone appears as a low pink-to-lavender cut on the north side of the road. The Deadwood formation, the oldest and lowest of the sedimentary layers, crops out poorly here. Watch for small outcrops of red and yellow Deadwood sandstone north of the road, near the Hisega turnoff. These are probably beach sands or coastal bars that formed as the Cambrian sea flooded eastward across South Dakota. They are the lowest bedded aquifers in the area.

Between Hisega and US 385, watch for the flat layers of the Deadwood sandstone lying on Precambrian rocks with steeply inclined layers. The Precambrian rocks are mostly quartzites and schists that began as sands and sandy clays laid down on an ancient seafloor. Some outcrops expose massive dark green rocks between the layers of sedimentary rocks. These are metamorphosed basalts and gabbros, originally lava flows or sills. Age dates suggest they are about 2.3 billion years old. None of these rocks contain metallic ore deposits, so they cannot be the source of the placer gold along Rapid Creek. It probably came from the veins near Rochford.

Big Bend

The Big Bend is an old meander of Rapid Creek that entrenched as the stream eroded its channel from an old floodplain some hundreds of feet above its present level.

About 1893, a Rapid City promoter decided to drive a tunnel across the neck of the Big Bend and divert Rapid Creek, thus drying up the 8,000-foot-long loop and making it accessible to placer miners. He sank a shaft 60 feet below creek level on the upper end of the loop and drove a tunnel 800 feet toward it from the lower end. The two did not meet quite as planned. They created a spectacular underground waterfall 30 feet high, now known as Thunderhead Falls. The amount of placer gold recovered in this caper was profoundly disappointing.

Johnson Siding

At Johnson Siding, on the long and sparsely vegetated slope to the northeast, the layers of Precambrian rocks stand nearly vertically beneath a rimrock of nearly horizontal Deadwood sandstone. The Cambrian sea advanced across an old erosion surface, depositing as much as 200 feet of Deadwood sands and shales over much of the area and finally burying the highest knobs of Precambrian quartzite.

Box Elder quartzite stands on end just east of Johnson Siding.

The Norris Peak road, Forest Service Road 166, goes north from here to join Nemo Road at the junction of Bogus Jim and Box Elder Creeks. Watch for good exposures of Deadwood sandstone at road level.

Placerville

Between 1900 and the start of World War II, Rapid Creek was again the focus of many placer-mining operations. Fine gold occurred in the gravels along the present stream and on terraces left high as the stream cut its canyon deeper. The miners called those high-level gravel deposits "bars." They needed water to work those dry gravels.

Hydraulic mining was the easiest way to wash the high gravels off the underlying bedrock surface, where the gold concentrated. This method involved jetting large volumes of water under high pressure against the face of the gravel bank. The miners washed the finer gravels and sands through sluice boxes to capture the gold. They washed or shoveled the coarse gravel to one side and stacked the boulders in heaps by hand.

The Pactola-Placerville Mining Company was established in 1894 to secure water. The company diverted Rapid Creek into a flume just below the present site of Pactola Dam. The flume followed the south side of the canyon at a gradient much less than that of the creek, thus raising the flume steadily higher above the creek. Where possible, the flume was an earthen trench, but it crossed tributary gullies in a rectangular timber trough.

Water was drawn from the flume at a succession of bars. At Placerville, the flume was only about 40 feet above the creek, so the water pressure was not high enough to be very effective. At the public fishing grounds at Placerville, you can still see the large piles of boulders and the denuded bedrock surface where the miners dug into crevices to recover the heavy flakes of gold. Old tunnels into the gravel bank are still open.

Hydraulic mining ended in 1906 or 1907. In 1907, the Dakota Power Company took over the flume and extended it to a point above Big Bend as a wooden-stave pipe 4 feet in diameter. There the water dropped 270 feet through a penstock to drive turbines in a small hydroelectric generating plant, which provided electricity to Rapid City between 1912 and World War II. You can still see notches along the canyon wall and occasional clusters of old timbers on the hillside above SD 44.

Maverick Junction–Rapid City

Were it not for remnants of badlands clays and for patches of recent alluvium in the valley bottoms, the highway between Maverick Junction and Rapid City would lie entirely on somber Cretaceous shales. The strong color contrast between the dark gray shales and the light-tan-to-pinkish clays of the White River group makes it easy to follow their distribution. Many patches of cemented gravels mark the courses of Oligocene streams that flowed from the Black Hills onto the desert alluvial plain on which the sediments of the White River group were accumulating.

Maverick Junction to Rapid City (51 miles).

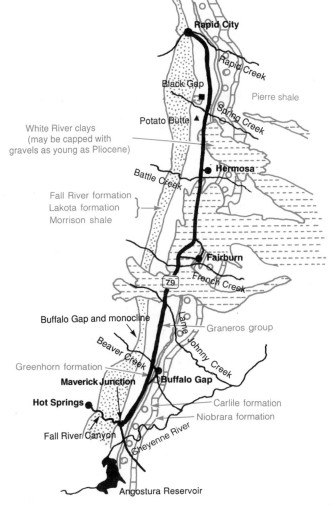

319

Thin veneers of Tertiary gravels, as at Black Gap, cap high terraces between the streams. Cretaceous rocks lie beneath them. Wells show that some of the valleys still contain several tens of feet of badlands clays. Evidently, the topographic relief was greater at the end of Eocene time, when those sediments began to accumulate, than it is now.

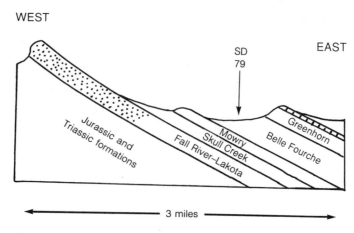

WEST

EAST

SD 79

Jurassic and Triassic formations

Fall River–Lakota

Skull Creek

Mowry

Belle Fourche

Greenhorn

← 3 miles →

Cross section just north of Maverick Junction. Pine trees grow on the sandstone and on the siliceous Mowry shale. Yucca grows on the limy Greenhorn shale.

Buffalo Gap Monocline

Beaver Creek departs the Black Hills through a water gap it eroded in the hogback ridge of Cretaceous formations. The sandstones are nearly flat on the plateau a mile to the west, but roll sharply down to the east and stand at an angle of 70 degrees where they disappear into the valley floor. This is one of several similar monoclines around the margin of the Black Hills. They formed as the sedimentary formations near the surface draped over a fault in the deeper rocks. In this case, the rocks east of the fault dropped several hundred feet.

Beaver Creek loses its flow to caverns within the Pahasapa, or Madison, limestone near the western edge of Wind Cave National Park. Some of the water returns to the surface through a spring just inside Buffalo Gap at the fish hatchery. Some may also find its way southward to Cascade Springs.

Buffalo Gap has apparently been a favorite access route to the Black Hills for both animals and man. The first written record is in the diary of one of the mountain men who accompanied Jedediah Smith in 1823. His description clearly identifies Buffalo Gap, but the rest of the journal gives little clue to their route through the Black

Hills. However, his mention of rocks that "glistened in the sunlight" almost certainly refers to the mica schists at the core of the range.

Greenhorn Escarpment

The steep tilt of the rock layers in the Buffalo Gap monocline narrows the outcrop belt of the lower Cretaceous shales, bringing the ridge of Greenhorn limestone close to the east side of the highway. Some 35 to 50 feet of slabby and coarsely crystalline limestone protect the crest of the escarpment from erosion. If you break a piece, you will smell the strong odor of petroleum.

The Greenhorn limestone is full of fossils. The most common are fragments of the big clam mytiloides, formerly known as inoceramus. It is generally in small pieces because its shells had a thick middle zone composed of long crystals of calcite lined up parallel to each other, like a fistful of pencils. They separated easily to make more or less rectangular fragments. In some places the Greenhorn limestone consists entirely of those tiny prisms of golden calcite broken from the middle layer of the clamshells. Those are one of the characteristic features that enable geologists to recognize sedimentary rocks laid down in shallow seawater during late Cretaceous time.

Beneath the slabby Greenhorn limestone is about 150 feet of light gray calcareous shale, also in the Greenhorn formation. It supports a generous stand of yucca, or Spanish bayonet. At the base of the gray shale member is a very thin sandy limestone with small shark teeth plastered on the bedding planes. This thin limestone separates the Greenhorn shale from the underlying and much darker shale of the Belle Fourche formation, which is not calcareous.

The accompanying geologic strip map shows the Graneros group— all strata between the forested sandstone ridge that forms the western skyline and the base of the Greenhorn formation. When N. H. Darton of the U.S. Geological Survey came from Colorado to map the Black Hills around 1900, he called this interval the Graneros formation because it resembles beds of about the same age in south-central Colorado. After more detailed study of this sequence, the original Graneros formation of the Black Hills was divided into four formations, which together make the Graneros group. The following terminology is now in use:

FORMATION	THICKNESS
Belle Fourche shale	600 feet
Mowry shale	150–200 feet
Newcastle sandstone	0–25 feet
Skull Creek shale	175–225 feet

Mowry Shale

At several points north and south of Buffalo Gap, the highway skirts knolls of gray and rather siliceous Mowry shale. They support a good growth of pine trees. In Wyoming, Colorado, and Montana, the Mowry shale also makes conspicuous forested ridges in otherwise treeless expanses of Cretaceous shale. Most of the Mowry shale on the eastern flank of the Black Hills is not siliceous; it weathers to a tight gumbo clay on which no pine trees grow. The more siliceous outcrops contain a generous abundance of small fish scales and bone fragments.

Calico Canyon

Calico Canyon, the first canyon south of Buffalo Gap, contains outcrops of Unkpapa sandstone, colorfully banded in shades of white, yellow, and purple. The many small faults offsetting the bedding planes probably developed as the sand compacted before it was ce-

Natural bridge in Unkpapa sandstone at head of Calico Canyon near Buffalo Gap.

mented into solid rock. Slabs of this attractive rock have been used as an interior decorative stone. The small natural bridge at the upper end of the canyon is within the Unkpapa sandstone.

Fairburn Agates

Terrace gravels associated with French Creek below Fairburn contain agates that were probably eroded from cherty zones within the Minnelusa formation farther upstream. The intricate patterns of al-

Cluster of polished Fairburn agate. —S.D. Museum of Geology

ternating layers of red, black, and white quartz have led many people to call them fortification agates. The red color is iron oxide; the black is manganese oxide. Fairburn agate is the state gem of South Dakota. You can see a choice collection at the Museum of Geology at the School of Mines at Rapid City.

Fairburn Anticline

Just northeast of Fairburn, but not visible from the highway, is a sharply folded anticline. An outcrop of Greenhorn limestone traces its oval map outline. A wildcat well drilled in 1949 found a small showing of high-grade crude oil, but not nearly enough to be of commercial interest.

A Distinct Lack of Uranium

A small sandstone ridge covered with trees just west of Hermosa and north of SD 40 exposes Newcastle sandstone that was deposited in a stream channel. Reported showings of uranium on the outcrop prompted several dozen test holes in the Hermosa area in the 1970s. They found no commercial ore body.

Potato Butte

Potato Butte stands over 100 feet above the present landscape. It is capped by a thin lens of freshwater algal limestone formed in a small, temporary pond on the old Oligocene plain.

Black Gap

A thick layer of solidly cemented gravel holds up the high terraces on both sides of the highway. The gravel was deposited during Oligocene time, obviously on an older erosion surface. The dark gray shale beneath is the Belle Fourche formation. Springs exist at low points along the contact between the porous gravel and the watertight shale beneath. You can spot them by looking for clumps of wild shrubs in the draws.

The highway passes through a notch that in 1877 was the route of the first stage and freight lines between the Black Hills and the Union Pacific Railroad at Sidney, Nebraska. Stage drivers rested their horses at the crest after the long pull up the muddy gumbo grade. On more than one occasion, robbers picked this point to rob stages leaving the Black Hills with gold dust and bullion destined for the U.S. Mint at Denver.

Experimental lignite coal gasification plant.

Free-swimming reptiles. —S. D. Museum of Geology

Gasification Plant

The federal government built the tall, skeletal structure on the south edge of Rapid City as an experimental plant to test the carbon dioxide acceptor process for converting coals into enriched pipeline gas. The plant heated a mixture of finely ground coal and dolomite under high pressure to produce gas with a much higher heating value than that obtained by heating the coal alone. The pilot plant opened in 1971 and operated for a few years at a capacity of 30 tons of coal per day. It tested lignite coals from North Dakota and Texas and higher grade coals from Montana and Wyoming. The plant closed with the completion of the testing program.

School of Mines

The South Dakota State School of Mines was established in 1886, before South Dakota became a state. It was founded to train miners in geology, mining, and metallurgy only during the winter months, but it rapidly evolved into an accredited school offering a variety of degrees in science and engineering. The Museum of Geology, established when the school opened, is in the Administration Building. It displays an exceptional collection of minerals and fossils and makes available study collections to visiting scholars. The scarred hill just behind the football field was the site of a large gold and silver smelter that operated briefly about 1900.

SD 87, Needles Highway
US 16–US 16A
20 miles

This short drive southeast on SD 87 from the junction with US 16 goes through spectacular mountain scenery in the central part of the Black Hills, passing outcrops of a wide variety of Precambrian rocks. The Precambrian rocks are part of the ancient continental basement, the very stuff of the continental crust itself, a complex of metamorphic rocks heavily injected by granite. These are among the most baffling and beautiful of all kinds of rocks. A close look at almost any specimen reveals large and glittering crystals, many of them colorful.

Sunday Gulch

Between Sylvan Lake and US 16 south of the Mount Rushmore turnoff, SD 87 leads up Sunday Gulch. The bedrock is gray mica schist intruded by countless granite dikes, some very large. The schist weathers more readily than the granite, so it rarely makes good outcrops. It underlies the valleys and notches between the bold ridges and spires of the granite dikes.

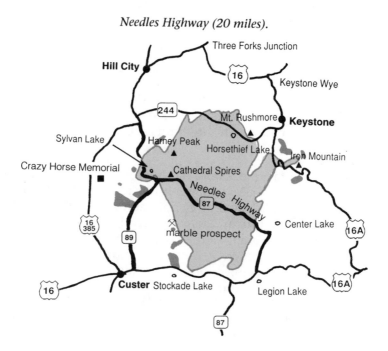

Needles Highway (20 miles).

Cathedral Spires from Needles Highway.

Blue Star Spring is just below the first switchback. It was a welcome spot for filling radiators in earlier years, and it is still a source of coffee water for local residents who object to chlorine in their municipal water. The spring flows slightly less than 5 gallons per minute at a temperature of 47 degrees Fahrenheit in all seasons. Most springs produce water at the mean annual temperature of the area.

Sylvan Lake

In 1892, a private group built a masonry dam across a narrow defile in the granite at the head of Sunday Gulch. At an elevation of 6,145 feet, Sylvan Lake is the highest in the state. In 1920 the state

Sylvan Lake. —S.D. Department of Tourism

purchased the lake and a resort hotel built in 1893. The present hotel was built in 1937, a year after the original one burned. Both lake and hotel are an integral part of Custer State Park.

The Needles

The most spectacular part of SD 87 lies between Sylvan Lake and the junction with US 16A near Legion Lake. The state built this stretch of road as a scenic route in 1920 and 1921 at a total cost of only $167,000. One reason for the modest cost was that all blasting was done with explosives left as surplus from World War I.

In 1979 the National Park Service designated a tract of 640 acres as a Registered Natural Landmark. It includes the Cathedral Spires, the Needles Eye, and the Little Devils Tower. The spires are an excellent example of how fractures control weathering of granite. In highway cuts, you can see both the change from fresh to weathered granite along fractures and the relative rates at which different minerals weather: deeply weathered feldspar, slightly altered mica, and fresh, angular quartz.

South Dakota's only known stand of limber pine is here, apparently a relic of the ice age flora. The trees cover about six acres in a protected valley between the Cathedral Spires and the Little Devils Tower. They are up to 200 years old and reach heights of 50 to 80 feet. The nearest stand of limber pine is in the Laramie Range, 150 miles to the southwest.

Granite pinnacles, Needles Highway.

Marble

About 1900, quarrymen tried to exploit a mass of Precambrian dolomitic marble for building stone. The rock grades from pure white to mottled white and gray; some layers contain flecks of green serpentine. Its poor accessibility and distance from a market doomed the project. The old quarry is now within Custer State Park.

Center Lake

Center Lake is a reservoir impounded behind a dam on Grace Coolidge Creek, formerly Squaw Creek. President Coolidge spent the summer of 1927 in the Black Hills during renovation of the White House, hence the new name. From Center Lake south to the junction with US 385, SD 87 crosses a monotonous sequence of rocks shown on geologic maps as Precambrian metaquartzite, which is sandstone deposited during Precambrian time, then metamorphosed to quartzite. It also includes light tan siliceous schist.

SD 89
Pringle–Minnekahta Junction
16 miles

This short link between US 18 at Minnekahta Junction and US 385 at Pringle crosses open country, passing excellent roadcuts through late Paleozoic carbonate rocks. At Minnekahta Junction,

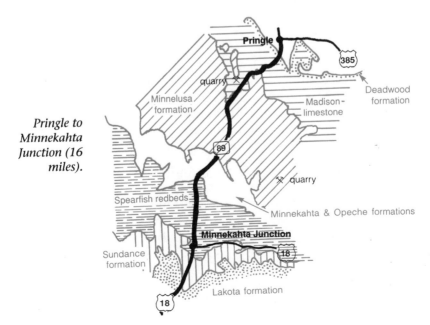

Pringle to Minnekahta Junction (16 miles).

outcrops of Permian and Triassic Spearfish redbeds form the bedrock in the Red Valley. The road drops through the following
section to the Precambrian schists at Pringle:

AGE	FORMATION	THICKNESS
TRIASSIC		
	Spearfish redbeds	350 feet
PERMIAN		
	Minnekahta limestone	35 feet
	Opeche redbeds	100 feet
PENNSYLVANIAN and PERMIAN		
	Minnelusa sandstone and dolomite	850 feet
MISSISSIPPIAN		
	Pahasapa, or Madison, limestone	350 feet
CAMBRIAN		
	Deadwood sandstone and shale	50 feet
PRECAMBRIAN		
	schist	

Limestone outcrops in the southern Black Hills furnish commercial rock to a large area of Nebraska and eastern Wyoming where all
rocks are soft Tertiary formations that contain nothing hard enough
for road metal or construction aggregate. One large quarry exploits
the exceptionally pure Minnekahta limestone; several others use the
Madison limestone. Quarry sites were opened along a spur line of
the railroad, which paralleled the old highway.

Rerouting of the highway opened many good roadcuts in the Madison and Minnelusa formations. Colorful outcrops are thin red shales
within the Minnelusa formation and a thick zone of residual red soil
at the top of the Madison limestone.

Reed Cave

Routine blasting at the big Madison limestone quarry at Loring
Siding exposed Reed Cave in 1966. It is a fair-sized cavern system.
Students from the School of Mines and others have mapped more
than 3 miles of accessible passages, all with a general northeast trend.
Reed Cave has not been developed for public visits.

Nemo Road, Forest Service Road 2335
(Pennington County Road 234)
Rapid City–US 385
38 miles

This scenic drive logically divides itself into two segments. An eastern leg is through canyons cut in the colorful Paleozoic sedimentary rocks that dip very gently down to the east. A western leg, west of Nemo, passes through a very complex Precambrian terrain of drab gray rocks, including massive quartzites, rocks of the banded-iron formation, conglomerates, phyllites, and a thick sill of what was originally a dark, igneous rock, gabbro.

Canyon walls adjacent to the highway spectacularly display the following sedimentary rocks. Read down if you are going west, up if you are driving toward Rapid City:

Minnekahta limestone and underlying Opeche red shale
- Permian
- 130 feet thick
- Seen only briefly at mouth of South Canyon

Minnelusa formation
- Pennsylvanian and Permian
- About 500 feet thick
- Red and yellow sandstones and dolomites; mostly sandstone in upper half, dolomites in lower half

Pahasapa, or Madison, formation
- Mississippian
- 450 feet thick
- Massive gray limestone, cavernous, cliff forming

Englewood limestone
- Latest Devonian to earliest Mississippian
- 30 feet thick
- Thin-bedded pink shaly limestone, grades into overlying Madison limestone

Deadwood formation
- Late Cambrian
- 200 to 400 feet thick
- Largely dark red sandstone, with some layers of dolomitic limestone and shale

Metamorphic rocks
- 2.6 to 2.15 billion years old, depending on unit
- Mostly dark gray
- Because of great differences in hardness of the various units, the old Precambrian erosion surface had a local relief of at least 200 feet when the sea invaded during Cambrian time, depositing the Deadwood sandstone.

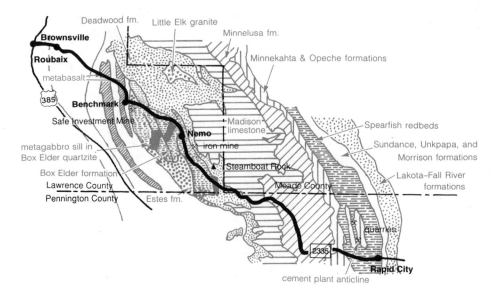

Rapid City to junction with US 385 (38 miles).

Rapid City

From downtown, drive west on Omaha Street through its merger with West Chicago Street and continue west for 2 miles. Then veer northwest on Nemo Road, Pennington County Road 234. The large battery of concrete silos north of West Chicago Street belongs to the cement plant. One mile brings you to the mouth of South Canyon. The bedrock throughout western Rapid City is redbeds in the Spearfish formation, but most are hidden under gravels of the floodplain of Rapid Creek or the higher terraces.

The road enters South Canyon through a notch in the Minnekahta limestone, which dips about 10 degrees down to the east. South Canyon is a deeply incised tributary of Rapid Creek. It exposes the limestone and sandstone units of the Minnelusa formation for about 4 miles. At the head of the canyon, the road emerges onto an upland surface developed on the divide between the Rapid Creek and Box Elder Creek drainages.

Box Elder Valley

From the Doty Fire Station, the road drops briefly into the canyon of Box Elder Creek. Excellent cliffs expose the Minnelusa formation. The 1972 flood eroded much of the topsoil from the valley floor. The roadbed going up the north side of the valley needs constant maintenance because it tends to slide.

Middle section of the Minnelusa formation. Sandstone grades down into well-bedded dolomite and sand.

Lone Grave

A field on the upland surface south of the road just east of the national forest boundary contains the grave of a soldier who died during the Custer expedition of 1874.

The contact between the Madison limestone and the overlying Minnelusa formation is near this short section of flat road. Watch for patches of red dirt. They are remnants of a residual soil that developed on the Madison limestone shortly after it was deposited, before a new invasion of the shallow sea during Pennsylvanian and Permian time laid down the Minnelusa formation.

Sinkhole Zone

The green bridge over Box Elder Creek, 1.2 miles west of the national forest boundary, is called the School Section Bridge.

Detailed map along part of Box Elder Creek.

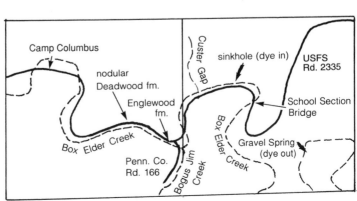

Water entering sinkhole in bed of Box Elder Creek below Custer Gap. —Ed Glassgow

In the half mile between its junction with Bogus Jim Creek and the School Section Bridge, Box Elder Creek flows below a cliff of Madison limestone and across a cavernous zone in the formation. Except during high water, the entire stream disappears through sinkholes into caverns in the limestone. Dye poured into one of these sinkholes just above the bridge reappeared in just over one hour in Gravel Spring, a half mile downstream. Below that spring for another half mile, the surface stream flow alternately disappeared and reappeared at successively smaller springs. The channel is normally dry for several miles below that point.

A third of a mile above the School Section Bridge, a steep ravine cuts the limestone bluff north of the creek. As Custer's wagon train

The dye appeared one hour and eight minutes later in Gravel Spring, a half air-mile downstream.

was leaving the Black Hills in August 1874, he had to decide whether to continue down the narrow and rocky canyon of Box Elder Creek or to manhandle his wagons up this ravine to gain the upland surface, which he could follow northeast toward Bear Butte and the open prairie. He chose the gap. Rope burns on rocks and trees were still visible a quarter of a century later when geologists first mapped the area.

Flood of 1972

Before they were cut to clear the channel, numerous pine trees along this stretch of Box Elder Creek preserved evidence of the flood of 1972. Two months after the flood, debris was still plastered against the trees 20 feet above normal creek level. Passing debris had ripped the bark from some of the trunks as much as 24 feet above the creek. Areas of pale gray limestone, exposed a few feet above the creek bed, show where the flood quarried and carried away the weathered rock surface, exposing fresh rock free of lichens.

Englewood Limestone

The highway curves sharply where Bogus Jim Creek and the Norris Peak Road, County Road 166, enter Box Elder Valley from the south. Watch for a nice pink outcrop of Englewood limestone on the inside of the bend in the road.

Sometime after the Ordovician sedimentary formations were deposited, this region rose above sea level and developed an erosional landscape covered with red soil. Then the land sank, or perhaps sea level rose, and the region was again shallowly submerged. The encroaching sea deposited the Englewood limestone as it advanced across the old erosion surface. The lowest several feet of the Englewood limestone owes its distinctly pink color to the red soil, which the waves worked into the accumulating limestone mud.

So the Englewood limestone is merely an impure basal phase of the thick Madison limestone. It would make more sense and be more convenient to include the Englewood limestone in the Madison limestone.

Concretions in the Deadwood Sandstone

A few hundred yards upstream from the outcrop of Englewood limestone, also north of the highway, is an outcrop of the upper part of the Deadwood sandstone. On the weathered surface it looks almost like a conglomerate of small sandstone cobbles. In fact, calcite cemented the sand grains around some kind of nuclei to make concretions. Those weather into bold relief, looking like sandstone cobbles. This is the only place in the Black Hills where this type of cementation is found in the upper Cambrian sandstone.

Irregular cementation in Deadwood sandstone makes it look like a conglomerate.

Steamboat Rock and Picnic Area

Steamboat Rock is an isolated remnant of the Madison formation that resembles a ship, if you have a knowing and properly initiated eye. It stands about 600 feet above Box Elder Creek. The Madison limestone caps the Deadwood formation, which rests on Precambrian talc at the base of the slope.

Watch at creek level for outcrops of darkly greenish igneous rock, the Precambrian Box Elder sill partly altered to talc. About 25 feet above the creek, a bench in the outcrop of the sill marks the roadbed of the old narrow-gauge railroad. Ten feet higher is the contact between the sill and the Deadwood sandstone, almost as flat as when it was originally deposited. The Deadwood sandstone continues 250 feel higher on the hillside.

Overlying it, but nearly buried under talus, is about 20 feet of green shale, the Winnipeg formation. It is a relic of Ordovician time. This exposure roughly marks the southern limit of Ordovician sedimentary rocks in the Black Hills. The sea probably extended farther south during Ordovician time, so the rocks deposited on its floor must have continued too. If so, they were lost to erosion before Mississippian time began.

Look above the green Winnipeg shale for about 30 feet of pink Englewood limestone, also very poorly exposed. Above that lies 250 to 300 feet of pale gray Madison limestone, which forms cliffs wherever it is exposed.

Box Elder Sill

A roadcut on a sharp curve 1 mile east of the Steamboat Rock picnic ground exposes a large outcrop of dark greenish gray rock. It is metamorphosed gabbro, part of a sill about 3,000 feet thick sandwiched within the Box Elder quartzite. The molten magma injected the Box Elder quartzite when it was still intact and lying flat. Both the quartzite and the sill were later folded and broken along large faults into pieces that appear in several places in the Nemo area. The sill must have had an original lateral extent of at least 10 miles.

Tiny crystals of zircon in the uppermost part of the gabbro sill have been dated at 2.15 billion years, so the quartzite is even older than that. Later, all of the rocks were folded and partially eroded. Vertical faults, offsetting the Box Elder quartzite, define basins where conglomerates formed from accumulating boulders and gravel of quartzite and banded-iron formation rocks. The Precambrian rocks were then folded so that the bottoms of the basins tilted past 90 degrees.

Strongly deformed Precambrian conglomerate is tilted on end and the pebbles are stretched.

The Lipalian Interval, an Abyss of Time

An outcrop 2 miles east of the Steamboat Rock picnic ground provides an exceptionally nice exposure of a profound unconformity between two sequences of sedimentary rocks. Park at the east end of the bridge over Box Elder Creek and look 150 feet south near the edge of the trees.

Look for layers of Box Elder quartzite tilted down at an angle of 70 degrees. An old erosion surface truncates those layers. Above it, and only slightly tilted, are layers of Deadwood sandstone. Age dates on the Box Elder sill show that the Box Elder quartzite, into which the sill was intruded, must be more than 2.15 billion years old, putting it in Precambrian time. The base of the Deadwood sandstone is late Cambrian, about 520 million years old. So the buried erosion surface between them must represent a gap of about 1.5 billion years.

Similar unconformities, representing approximately the same time interval, are widely exposed on all continents. Years ago, geologists wondered mightily about that terrific span of time, calling it the Lipalian interval. Hardly anyone has uttered that phrase for decades now, and geologists rarely discuss the time interval. Even so, the problem is as unsolved now as it was then.

Homestake Mining Company started timbering operations around Nemo in 1898, shipping lumber and mine timbers to Lead on their Black Hills and Fort Pierre Railroad. They built a modern mill using band saws at Nemo in 1914. After the railroad was abandoned in 1930, lumbering operations shifted to the western side of the Black Hills. Watch at the bend in the highway at the north edge of town for a nice exposure of yellowish brown Deadwood sandstone.

Banded Iron Formation

An irregular belt of banded-iron ore trends northwest from a point a couple of miles southeast of Nemo and continues at least 5 miles farther northwest. The formation closely resembles some of the iron formations in northern Minnesota. It consists of thin alternating bands of steely gray hematite and pale chert. The beds were originally laid down flat on an ancient seafloor; they are now intricately folded, overturned in places, and broken along faults.

Banded iron formation was deposited in many parts of the world back in Precambrian time, invariably before about 2 billion years

Steeply dipping Box Elder quartzite truncated by erosion and overlain by flat Deadwood quartzite.

ago. Most geologists believe that the oceans contained dissolved iron then, which reacted with oxygen that primitive algae released into the atmosphere. Iron oxide accumulated until all the iron dissolved in the oceans was used up. Only then could free oxygen begin to accumulate in the atmosphere.

A small open-pit iron mine sporadically produces hematite, or red iron oxide, from the iron formation. It is about a mile south of Nemo in the edge of the trees across the valley from the road. The hematite is hauled to Rapid City, where it is blended with other raw materials to make a special portland cement. The iron ore outcrop continues northwest from the mine and crosses the highway about a quarter of a mile south of Nemo.

Forest Service Road 135

Forest Service Road 135 leads ultimately to Sturgis. One side road from it leads to outcrops of the Little Elk Creek granite in the Dalton Recreation Area. The granite, actually a granite gneiss, is dated at 2.56 billion years, making it the oldest known rock in the Black Hills. Wonderland Cave is a commercially operated cavern in the Madison formation.

Safe Investment Gold Mine

From 1898 until 1930, Benchmark was a station on the old narrow-gauge line. A sharp curve in the highway marks the station site. North of Benchmark, the highway leaves the broad valley of Box Elder Creek, follows Hay Creek to its headwaters, then crosses the divide into the Elk Creek drainage.

A mile east of Benchmark, an inconspicuous pair of wheel tracks leaves the highway and leads to the ruins of the Safe Investment Mine. The dump where miners piled waste rock is on the south side of Box Elder Creek, partly overgrown. Nearby is the collar of a shaft, now caved in, but said to be 800 feet deep. Several old adits enter the slope horizontally about 80 feet above the creek. One leads to open stopes overhead, showing that it did produce some ore from metamorphosed conglomerate that contains pyrite.

High on a hill east and north of the highway is a large open pit from which aerial trams carried ore across the road to the mill. The gold was in an ancient placer deposit. It resembles some of the much larger and much richer gold deposits of the Witwatersrand in South Africa. A small stamp mill reportedly crushed ore as early as 1883. The operators moved a larger mill onto the property in 1906. No record of later mining seems to survive, and neither does any record of production.

Windy Flats

The road crosses slightly metamorphosed basalt flows, black shale, and muddy sandstone, all just below the old Precambrian erosion surface. An abandoned road-metal quarry north of the highway produced crushed rhyolite from a sill.

Roubaix

Prospectors discovered the ore body of the Uncle Sam Mine in 1878. Mining began soon thereafter and continued until a heavy flow of water closed the mine in 1889. In 1898, Pierre Wibaux bought the property. He was a Frenchman who came to eastern Montana and made a fortune in ranching and banking. He changed the name of the railroad siding from Perry to Roubaix after his hometown in France, and he renamed the mine the Cloverleaf. The mine and mill closed in 1905, after recovering nearly 43,000 ounces of gold.

The gold was in a mass of Precambrian quartz shaped like a saddle. It was on the crest of an anticlinal arch whose axis plunged down to the southwest at an angle of 30 degrees. Lateral drifts led from the shaft to the crest of the ore body at eight levels, the deepest of which was 705 feet below the surface. Coarse, metallic gold speckled the white quartz. Associated minerals were galena, sphalerite, and pyrite.

A stamp mill crushed the ore. It was then washed over a mercury-plated copper sheet where the gold amalgamated with the mercury and the other minerals went to waste. The amalgam was periodically scraped off the copper plate, placed in a retort, and heated. This drove the mercury off as a vapor, leaving the gold. The mercury vapor was run through a cooling coil to condense it for reuse. The property briefly reopened as the Anaconda Mine in 1936–37, when it produced an additional 300 ounces of gold and a small amount of silver.

Brownsville

Brownsville was originally a stop on the Black Hills and Fort Pierre Railroad. In 1881, seven years before outside railroads reached the Black Hills, Homestake Mining Company started building this narrow-gauge line south from Lead. It brought in wood for mine timbers and fuel for the steam engines that drove the mines and mills.

Between Brownsville and Nemo, the highway generally parallels the old narrow-gauge line. In many places, you can see the old grade skirting the trees at the edge of the valley of Hay Creek or Box Elder Creek. The Burlington Railroad took over the line in 1901, operated it until 1930, then abandoned it.

Supplemental Reading

Darton, N. H., and S. Paige. 1925. *Central Black Hills, South Dakota.* Geologic Atlas of the United States, folio no. 219. Reston, Va.: U.S. Geological Survey.

DeWitt, E., J. A. Redden, D. Busche, and A. B. Wilson. 1989. *Geologic Map of the Black Hills Area, South Dakota and Wyoming.* Miscellaneous Investigations Series, map I-1910. Reston, Va.: U.S. Geological Survey.

Feldmann, R. M., and R. A. Heimlich. 1980. *The Black Hills.* Dubuque, Iowa: Kendall/Hunt Publishing Co.

Froiland, S. G. 1978. *Natural History of the Black Hills.* Sioux Falls, S.D.: Center for Western Studies, Augustana College.

Harksen, J. C., and J. R. Macdonald. 1969. *Guidebook to the Major Cenozoic Deposits of Southwestern South Dakota.* Guidebook Series, no. 2. Vermillion, S.D.: S.D. Geological Survey.

Jennewein, J. L., and J. Boorman, eds. 1961. *Dakota Panorama.* See especially chapter 2, "The Face of the Territory." Mitchell, S.D.: Dakota Territorial Centennial Commission.

Lisenbee, A. L. 1985. *Tectonic Map of the Black Hills Uplift, Montana, Wyoming, and South Dakota.* Map Series, no. 13. Laramie, Wyo.: Geological Survey of Wyoming.

Martin, J. E., ed. 1985. *Fossiliferous Cenozoic Deposits of Western South Dakota and Northwestern Nebraska.* Dakoterra Series, vol. 2, pt. 2. Rapid City, S.D.: South Dakota School of Mines and Technology.

O'Harra, C. C. 1920. *The White River Badlands.* Bulletin no. 13. Rapid City, S.D.: South Dakota School of Mines. Wall, S.D.: Wall Drug Store, 1976.

Paterson, C. J. and J. G. Kirchner. 1996. *Guidebook to the Geology of the Black Hills.* Bulletin no. 19. Rapid City, S.D.: South Dakota School of Mines and Technology.

Rich, F. J., ed. 1985. *Geology of the Black Hills, South Dakota and Wyoming,* 2nd ed. Field Trip Guidebook Series, Geological Society of America. Alexandria, Va.: American Geological Institute.

Roberts, W. L., and G. Rapp, Jr. 1965. *Mineralogy of the Black Hills.* Bulletin no. 18. Rapid City, S.D.: South Dakota School of Mines and Technology.

Rothrock, E. P. 1943. *A Geology of South Dakota.* Bulletin no. 13. Vermillion, S.D.: S.D. Geological Survey.

Glossary

Adit—A gently inclined entrance to a mine. You can walk into an adit.

Amalgamation—A technique for recovering finely divided gold by combining it with mercury, then distilling off the mercury to recover the gold.

Anhydrite—A calcium sulfate mineral, which can make up a bedded rock when precipitated from an evaporating body of seawater.

Anticline—An arch folded in layered rocks.

Aquifer—A body of rock porous enough to hold substantial amounts of water and permeable enough to deliver it to a well. Well drillers bore in hopes of finding an aquifer.

Artesian water—Water from an artesian well. Except for being under pressure in the aquifer, artesian water is no different from any other water.

Artesian well—A well that produces water under pressure so that it rises in the well bore, sometimes high enough to flow at the surface.

Badlands—Extremely rough topography, developed in soft sediments on a small scale, with steep slopes, many drainages, and narrow canyons.

Bed or bedding—Individual layers within a series of sedimentary rocks. The bedding plane is normally parallel to the surface on which the sediment was deposited.

Bedrock—Solid rock in its original place.

Bentonite—A soft yellow clay consisting primarily of the clay mineral montmorillonite. It is capable of absorbing large volumes of water and is useful for sealing dams, foundations, etc. It originally fell as volcanic ash and later became clay.

Boulder—A rounded rock with a mean diameter greater than 10 inches.

Box canyon—A canyon with steep-sided walls, closed at the upper end. It typically forms where a resistant rock layer caps softer, more erodible strata.

Breaks—The deeply dissected country next to a major river valley.

Breccia—A rock composed of large, angular fragments cemented into a solid mass.

Butte—A small, isolated hill, typically an erosional remnant.

Calcite—The mineral form of calcium carbonate. Calcite is the principal constituent of limestone, dripstone in caves, caliche in soil, and boiler scale in a teakettle.

Cave or cavern—A large opening eroded in limestone as it dissolved in groundwater.

Chalk—A soft limestone typically composed of the microscopic shells of floating organisms. It typically occurs in sections of late Cretaceous sedimentary rocks.

Chert—A microcrystalline variety of quartz with a conchoidal fracture. It is most familiar as the typical material of stone arrowheads.

Claystone—An extremely fine-grained sedimentary rock; lithified clay.

Cobble—A rounded rock in the size range from a tennis ball to a volleyball.

Concretion—A hard, often rounded mass formed in a sedimentary rock by precipitation from groundwater around some type of nucleus, commonly a fossil. Those in South Dakota are generally composed of limestone, silica, or pyrite.

Cutbank—A steep, bare stream bank.

Dike—A tabular mass of igneous rock formed as molten magma injects a fracture in older rock.

Dip—The maximum slope of a tilted layer of rock.

Dip slope—The topographic surface of a tilted layer of rock.

Disconformity—An erosional break within a stratigraphic series strata, with beds above being parallel to those below.

Dolomite—1. A mineral similar to calcite except that one molecule of calcium is replaced by one of magnesium. 2. A sedimentary rock made largely of the mineral. Dolomites and limestone look very much alike.

Doré—A alloy of gold and silver.

Drift—1. A collective term for all materials deposited by a glacier or by meltwater from a glacier. 2. Mine workings that branch horizontally from a shaft or an adit.

Esker—A long sinuous ridge of gravel deposited by a stream flowing beneath a glacier.

Fault—A fracture within the earth's crust, along which one side has moved relative to the other.

Feldspar—A family of aluminum silicate minerals abundant in igneous and metamorphic rocks. Feldspar typically occurs as blocky crystals that are white, greenish white, or pink. It is the most abundant mineral group in the earth's crust.

Flint—Black chert.

Floodplain—The area near a stream that is submerged during periods of high water.

Formation—A recognizable rock unit large enough that geologists can plot its boundaries and distribution on a map. Formations are customarily named after geographic localities where they were first studied or where they are typically exposed.

Glacial erratic—A rock transported by a glacier.

Glacial till—Glacial debris deposited directly from the ice. It is typically a mass of unsorted debris, with rocks of all sizes and shapes embedded in clay.

Glacier—A large mass of ice, formed by the accumulation and compaction of snow, which moves outward or downslope under its own weight.

Granite—A crystalline igneous rock consisting largely of feldspar, quartz, and mica.

Gravel—An accumulation of rounded, usually waterworn pebbles ranging in size from that of peas to that of eggs.

Graywacke—A sandstonelike rock consisting mainly of quartz, but containing considerable feldspar, rock fragments, and clay.

Groundmass—The finely crystalline part of an igneous rock, in which larger crystals may appear. Also called matrix.

Gumbo—A popular name for any sticky soil. In South Dakota, gumbo forms through weathering of shale or bentonite.

Gypsum—The mineral form of hydrated calcium sulfate. Varieties include transparent crystals of selenite, fibrous veins of satin spar, and massive alabaster.

Hogback—A steep, elongate ridge, protected from erosion by a steeply dipping resistant stratum.

Ironstone—A hard, fine-grained, often brown rock that consists mainly of clay cemented with iron oxide.

Laccolith—An igneous intrusion that has the form of a blister within a sequence of sedimentary rocks.

Limestone—A common sedimentary rock consisting primarily of calcite. It generally formed on a warm sea bottom and commonly contains fossils of marine shellfish.

Loess—Windblown silt. It consists largely of rock flour picked up by the wind in front of a melting glacier and redeposited downwind.

Marine sediments—Sediments deposited on the seafloor, primarily forming shale, limestone, and sandstone.

Meander—A loop or series of loops in the channel of a stream, typically those flowing on a very flat gradient.

Marl—Clay with a high lime content.

Matrix—Fine-grained material in which larger rock fragments or crystals may occur.

Member—A division of a geologic formation that has a distinct character, but only local extent.

Metamorphism—Recrystallization of a rock at high temperature and pressure.

Meteorite—A fragment of rock or metal that enters the earth from outside its atmosphere. Most probably come from the asteroid belt between Mars and Jupiter.

Mica—A family of silicate minerals, all of which tend to split into thin sheets.

Monocline—A fold in layered rocks that consists of a simple flexure, not an arch or trough. Most monoclines probably form when rocks drape over a fault at depth.

Moraine—Any accumulation of unsorted debris, or glacial till, deposited directly from melting glacial ice. It includes ridges of debris that accumulate along the margins of a glacier. Ground moraine is glacial till that has spread rather uniformly over the surface.

Nodule—A small mass of rock enclosed within less-resistant rock. Nodules tend to stand out in relief on a weathered outcrop and typically have formed due to differences in cementation.

Ore—Rock from which one or more valuable minerals may be extracted at a profit.

Outcrop—An exposure of bedrock not obscured by surface material.

Outwash—Water and glacial debris issuing from the margins of a melting glacier.

Outwash plain—A blanket of glacial silt, sand, gravel, and boulders dropped by outwash streams.

Pegmatite—Very coarse-grained granite that occurs either as veins within a granite mass or in the country rock outside the granite. Pegmatite may be tabular, as dikes, or may be of any shape. It is abundant in the southern Black Hills around Harney Peak.

Petrifaction—The process in which mineral matter is precipitated within buried organic matter. Many petrifactions faithfully preserve the original organic structure.

Phyllite—A metamorphic rock intermediate between slate and mica schist.

Pleistocene—The last 2 million years or so, during which a series of ice ages have come and gone.

Porphyry—A fine-grained igneous rock with larger crystals of one mineral embedded in the groundmass. It is common in the northern Black Hills.

Pothole—A small glacial lake that occupies a basin that formed when a mass of ice buried in glacial debris finally melted.

Quartz—The mineral form of silica, or silicon dioxide. Quartz is one of the commonest and most abundant minerals. It occurs in many kinds of rocks and in many varieties.

Quartzite—A sandstone in which quartz cements the sand grains so tightly that the rock breaks across, rather than around, the individual sand grains.

Redbeds—Sandstone, siltstone, and shale with a strong red color imparted by ferric iron oxide.

Riprap—Large blocks of broken rock placed on an earthen surface, such as the face of a dam, to prevent erosion.

Sandstone—A deposit of sand hardened into more or less solid rock.

Sanidine—The high-temperature variety of potassium feldspar found in volcanic rocks.

Schist—A metamorphic rock that splits readily in one direction. Schists may be named for minerals that are present as in quartz-mica schist or garnet schist.

Shaft—A steeply inclined entrance to a mine. You can fall into a shaft.

Shale—Mud or clay hardened into more or less solid rock.

Sill—A tabular mass of igneous rock that formed as molten magma injected between layers of sedimentary rocks.

Sinkhole—Holes that open at the surface as the roofs of caverns collapse. Sinkholes are common where the bedrock is limestone or gypsum.

Stope—A large underground chamber opened where ore was removed from a mine.

Syncline—A trough folded into layered rocks.

Table—A broad hill with a flat top.

Terrace—A flat topographic surface, typically a remnant of an old floodplain, left as a stream entrenches its valley floor to a lower level.

Unconformity—An erosional break within a series of sedimentary layers.

Vein—A tabular mineral filling within a fracture in older rock. It is most commonly quartz or calcite, but it may include valuable ore minerals.

Volcanic ash—Fine particles of volcanic glass blown into the air during an eruption. Ash may drift downwind for long distances before settling.

Water gap—A narrow valley eroded through a ridge of resistant rock.

Water table—The level at which the ground becomes water saturated. You see the water table exposed in lakes, in ponds, and in water standing in a well.

Index

Aberdeen, 43–45, 74–75
agates: fortification, 271; Fairburn, 271, 322; moss, 143
Alexandria, 30, 34–35
Altamont glacial stage, 17, 35, 41–42, 49, 64, 88; moraine, 27, 30, 65, 68–69
Alzadasaurus pembertoni, 102
American Island, 101
ammonites, 94
Andes, Lake, 74
Andover, 44
Annie Creek, 252
anticline: Belle Fourche, 131; in Precambrian rocks, 264–65; La Flamme, 243; Oglala, 125–26; Chilson, 286; Whitewood, 240; Fairburn, 323; Dudley, 281; Cascade, 284–86; Cottonwood, 288
Arikaree formation, 98, 155
Armour, 74
arsenic, 274, 278
Artesian (town), 80
artesian, aquifer, 8–9; springs from, 244, 298; wells in, 38, 70, 80–81, 83, 104–5, 116, 142, 160, 188–89
Ash, Ben, 143
Ashton, 76
Atlantosaurus copei, 134
Avon, 90–91

Bad River, 19, 114
Badlands National Park, 170, 197
Badlands Wall, 105, 116, 197
Badlands, Big, 170–72, 191–202
Bald Mountain, 70
banded iron formation, 215, 339
barosaurus, 234–35
Battle Creek, 19
Battle of the Little Bighorn, 37, 152, 163
Bear Butte Creek, 163, 245
Bear Butte, 162–64, 187–89
Bear Mountain, 215
Beecher Rock, 303–4
Belle Fourche River, 19, 157–58, 162
Belle Fourche shale, 313, 321
Belle Fourche, 131, 145–46, 148
Belvidere, 104, 175
Bemis glacial stage, 17, 41–42; moraine, 26–27, 68

Benchmark, 340
bentonite, 37, 100, 146–48
Beresford, 24
Beulah, Wyo., 244
Big Badlands. *See* Badlands
Big Bend Dam, 84
Big Bend: of Missouri River, 84–86; of Rapid Creek, 317
Big Foot Pass, 201
Big Sioux River, 22–26, 30–31, 43, 48, 56, 68; ancestral, 78
Big Stone City, 38, 40–41
Big Stone glacial stage, 17, 29, 41
Big Stone Lake, 38
Bijou Hills, 21, 118
Bijou quartzite, 118, 121
Billsburg, 160, 183
Bison, 151
Bitter Lake, 43
Black Gap, 324
Black Hills: anomalous location of, 217; caves in, 259–60; drainage pattern of, 217–19; exploration of, 203; mining in, 224–31; rise of, 173, 191, 214–17, 222, 244
Black Pipe Creek, 175
Blacktail Gulch, 249
Blue Blanket Lake, 46
Blue Dog Lake, 43
Blue Star Spring, 327
Blunt, 52, 66
Bob Ingersoll Mine, 275
Bonesteel, 121
Boulder Canyon, 245
Boulder Park basin, 247
Bowdle Hills, 21
Bowdle, 45–46
Box Elder Creek, 19, 233, 308, 332–34, 337–38
Box Elder quartzite, 215, 317, 338–39
Box Elder sill, 337
Box Elder, 107
box canyons, 33
Brandon, 33–34
Bridal Veil Falls, 252
Bristol, 43–44
Brookings till plain, 17, 26, 65, 78
Brookings, 48
Brownsville, 309, 341

About the Author

John Paul Gries began teaching geology at South Dakota School of Mines in 1936 and continues his work as a professor emeritus and consulting geologist. He has written 92 publications and technical reports and has been honored for his extensive studies in groundwater. Gries lives in Rapid City, South Dakota.